Praise for *Inside th...*

"What C-Level executives read to keep their decisions. Timeless classics for indispensable Costello, Manager-Corporate Marketing Communication, General Electric (NYSE: GE)

"Want to know what the real leaders are thinking about now? It's in here." - Carl Ledbetter, SVP & CTO, Novell, Inc.

"Priceless wisdom from experts at applying technology in support of business objectives." - Frank Campagnoni, CTO, GE Global Exchange Services

"Unique insights into the way the experts think and the lessons they've learned from experience." - MT Rainey, Co-CEO, Young & Rubicam/Rainey Kelly Campbell Roalfe

"Unlike any other business book." - Bruce Keller, Partner, Debevoise & Plimpton

"The Inside the Minds series is a valuable probe into the thought, perspectives, and techniques of accomplished professionals. By taking a 50,000 foot view, the authors place their endeavors in a context rarely gleaned from text books or treatise." - Chuck Birenbaum, Partner, Thelen Reid & Priest

"A must read for anyone in the industry." - Dr. Chuck Lucier, Chief Growth Officer, Booz-Allen & Hamilton

"A must read for those who manage at the intersection of business and technology." - Frank Roney, General Manager, IBM

"A great way to see across the changing marketing landscape at a time of significant innovation." - David Kenny, Chairman & CEO, Digitas

"An incredible resource of information to help you develop outside-the-box..." - Rich Jernstedt, CEO, Golin/Harris International

"A snapshot of everything you need..." - Charles Koob, Co-Head of Litigation Department, Simpson Thacher & Bartlet

www.Aspatore.com

Aspatore Books is the largest and most exclusive publisher of C-Level executives (CEO, CFO, CTO, CMO, Partner) from the world's most respected companies. Aspatore annually publishes C-Level executives from over half the Global 500, top 250 professional services firms, law firms (MPs/Chairs), and other leading companies of all sizes. By focusing on publishing only C-Level executives, Aspatore provides professionals of all levels with proven business intelligence from industry insiders, rather than relying on the knowledge of unknown authors and analysts. Aspatore Books is committed to publishing a highly innovative line of business books, redefining and expanding the meaning of such books as indispensable resources for professionals of all levels. In addition to individual best-selling business titles, Aspatore Books publishes the following unique lines of business books: Inside the Minds, Business Bibles, Bigwig Briefs, C-Level Business Review (Quarterly), Book Binders, ExecRecs, and The C-Level Test, innovative resources for all professionals. Aspatore is a privately held company headquartered in Boston, Massachusetts, with employees around the world.

Inside the Minds

The critically acclaimed *Inside the Minds* series provides readers of all levels with proven business intelligence from C-Level executives (CEO, CFO, CTO, CMO, Partner) from the world's most respected companies. Each chapter is comparable to a white paper or essay and is a future-oriented look at where an industry/profession/topic is heading and the most important issues for future success. Each author has been carefully chosen through an exhaustive selection process by the *Inside the Minds* editorial board to write a chapter for this book. *Inside the Minds* was conceived in order to give readers actual insights into the leading minds of business executives worldwide. Because so few books or other publications are actually written by executives in industry, *Inside the Minds* presents an unprecedented look at various industries and professions never before available.

INSIDE THE MINDS

Inside the Minds:
Security Matters
Industry Leaders on Protecting Your Most Valuable Assets

If you are interested in forming a business partnership with Aspatore or licensing the content in this book (for publications, web sites, educational materials), purchasing bulk copies for your team/company with your company logo, or for sponsorship, promotions or advertising opportunities, please e-mail store@aspatore.com or call toll free 1-866-Aspatore.

Published by Aspatore Books, Inc.

For corrections, company/title updates, comments or any other inquiries please email info@aspatore.com.

First Printing, 2004
10 9 8 7 6 5 4 3 2 1

Copyright © 2004 by Aspatore Books, Inc. All rights reserved. Printed in the United States of America. No part of this publication may be reproduced or distributed in any form or by any means, or stored in a database or retrieval system, except as permitted under Sections 107 or 108 of the United States Copyright Act, without prior written permission of the publisher.

ISBN 1-58762-144-4 Library of Congress Control Number: 2003116934

Inside the Minds Managing Editor, Laura Kearns, Edited by Jo Alice Hughes, Proofread by Eddie Fournier, Cover design by Scott Rattray & Zoey London

Material in this book is for educational purposes only. This book is sold with the understanding that neither any of the authors or the publisher is engaged in rendering medical, legal, accounting, investment, or any other professional service. For legal advice, please consult your personal lawyer.

This book is printed on acid free paper.

A special thanks to all the individuals who made this book possible.

The views expressed by the individuals in this book (or the individuals on the cover) do not necessarily reflect the views shared by the companies they are employed by (or the companies mentioned in this book). The employment status and affiliations of authors with the companies referenced are subject to change.

If you are a C-Level executive interested in submitting a manuscript to the Aspatore editorial board, please email jason@aspatore.com. Include your book idea, your biography, and any additional pertinent information.

Inside the Minds:
Security Matters
Industry Leaders on Protecting Your Most Valuable Assets

CONTENTS

Howard A. Schmidt   7
BUILDING A MOSAIC OF SECURITY FOR A BETTER WORLD

William Boni   29
PROTECTION OF THE INTANGIBLE: THE 21ST CENTURY SECURITY PARADIGM

Margaret E. Grayson   55
SECURITY FOR A CONNECTED WORLD: IT'S ALL ABOUT TRUST

Bruce Davis   77
EMERGING SECURITY RISKS AND SOLUTIONS IN THE DIGITAL ERA

Kurt Long   89
IMPROVE SECURITY WITH IDENTITY MANAGEMENT

Phil Libin   103
A VERY BRIEF OVERVIEW OF ACTIVE SECURITY

Firas Bushnaq 121
LEARNING TO ADOPT SECURITY PROACTIVELY

Joseph W. Pollard 139
SOLVING PROBLEMS FOR CUSTOMERS: THE PATH TO CONTINUED GROWTH

Thomas Noonan 157
SIMPLIFYING SECURITY FROM THE INSIDE OUT

Christopher Zannetos 173
BALANCING RISK, COST, AND SERVICE QUALITY IN INFORMATION SECURITY

William Saito 189
LEARNING AS WE GO: SECURITY GROWS UP

Eric Pulaski 207
PROTECTING BUSINESS-CRITICAL IT INFRASTRUCTURES: EVALUATING IT RISK MANAGEMENT AND SECURITY

Building a Mosaic of Security for a Better World

Howard A. Schmidt, CISSP, CISM
eBay Inc.
VP Chief Information Security Officer;
Former White House Cyber Security Advisor and
Chief Security Officer Microsoft Corporation

Build in Security Early

We have entered a new era where threats and vulnerabilities have a dramatic impact on the ability to survive in business today. Security is part of the core business process for any online activity. Whether for e-commerce or email, whether for the information technology infrastructure to support brick-and-mortar activity or a telecommunications backbone – security must be built-in.

As we look into the future, my vision for where the industry is heading is threefold. First and foremost, computer systems will evolve to become self-repairing and self-healing so a business doesn't require a tremendous quantity of resources to maintain its systems.

Second, there will be an environment where quality control and engineering are built into the development of hardware and software. Many of the current vulnerabilities are not the result of someone's learning how to break something, but of flaws built into the systems that are not caught.

Third, we will be strong on multiple-factor authentication. Smart cards and biometric security devices will become as common as ATM cards are today.

Enterprise Approach to Security Issues

Many companies today have their entire existence in an online world. They depend on all aspects of the critical infrastructure – power, transportation, telecommunications, and the Internet. We have an enormous dependency on security, not only for providing security for the online community, but also for ensuring that everyone else has security, so we can receive the infrastructure support to keep business running.

Some people look at security as only the products that have a specific security-related function, such as items for personal use, anti-virus protection, personal firewalls, and passwords. But it is important to look at *all* applications that run on computer systems. In the broad sense of security, all products *must* have code developed for security.

The classic example of this imperative is a program that plays music on your computer. Many people don't think much about security when they are playing a CD. But if an application installed on your computer has a security flaw that gives someone unauthorized access to your computer, you have a security issue.

In many companies, approximately 20 percent of the security team's time is spent evaluating various products that support specific security functions, such as personal firewalls, antivirus software, strong authentication, and Web services. This intense evaluation is performed to limit the complexity of the overall system while maintaining a secure environment. During the evaluation process, the team needs to meet frequently with vendors in an effort to make a good, secure system even better for the next generation.

Many companies are very proud of conceiving new security ideas and actually filing for patents on them. Security teams should talk with vendors of all sizes, from major international companies with recognized reputations to small start-ups with good ideas. Neither the size nor the experience of a company should be the determining factor in whether the company has a worthwhile product. The bigger issue is the problems solved by the product.

The biggest mistake any of us can make is to automatically assume that a question we may have is already answered and there is something out there to deal with it. It is always valuable to spend time looking at available solutions. Doing so helps us understand not only what this generation is doing, but also what can be done for the next generation of security products and services.

In terms of research and development, it is important that we leverage R&D from other sources, such as universities and other IT companies. Often it makes little or no sense to "build your own" when R&D has moved capabilities so much further ahead.

Avoiding Breaches, Protecting Privacy and Intellectual Data

One of the greatest needs for security is to protect privacy. There are two facets to privacy. One relates to data protection, where a company works to secure its systems to make sure unauthorized individuals cannot gain access. Without security, privacy does not exist.

The second aspect involves control over confidential information. Security offers the ability to decide who gets to see information provided to a particular company and what the company does with it. As the provider, you must always have a clear understanding of what you will allow a company to do with your information, as opposed to what that company wants to do with it.

Setting up clear, intelligible guidelines in common, everyday language is crucial to ensuring that privacy is met in all aspects of security.

Individuals need to recognize and be aware that they are part of a bigger environment. Most of us have grown up knowing we should lock our valuables in a safe and keep our car doors and houses locked. We take all those independent actions to protect our individual assets and our personal property.

But the online world is totally different. For example, in the recent past, unsecured home computers have been used to launch attacks against other computer systems. People have had their personal information stolen – in crimes known as identity theft – on insecure individual systems. And hackers have used individual systems to connect to business systems, corporate networks, and remote access, opening gateways into corporate environments.

Securing your personal environment is a much greater necessity now because you are part of that larger collective.

Easy Breaching Points: How to Stay Ahead of Hackers

Just as cars require regular maintenance, like tune-ups and brake checks, similar maintenance must be performed as part of a day-to-day routine in the ecology of IT systems. An important way to protect systems from hacker intrusions is to stay current with patches and maintenance. Statistics from the Department of Defense show that more than 98 percent of the breaches of Defense systems each year did not result from a newly found exploit, but from negligence: The breached systems were not current with patches.

When you examine your systems environment, try to see it as you would see life in the physical world. Do what you can to protect the systems by making sure that you have intrusion detection and prevention, that you have alarms that activate when anomalies occur, and that you have a mechanism in place that provides an electronic trail when you think there is a compromise to confirm whether an intrusion has occurred.

I think the major protection, if all those fail, is that you have a response mechanism in place. You have specialized, trained personnel to deal with the breaches, to determine how to stop them, to identify what has happened during the breach, and to prevent the same kind of intrusion in the future.

Online Transaction Assurance

Generally, a process flow occurs during an online transaction. It starts when a person turns on a computer, connects to the Internet, and then uses this connection to authenticate to an e-commerce site.

The success of these steps is based on how securely an identity is protected during each phase and how secure the site is for transferring goods successfully with respect to privacy and availability. Customers should feel that, whether it is 2:00 in the morning or 2:00 in the afternoon, they have the ability to complete an entire transaction.

If a company's environment is fairly constant, where transactions are very cut-and-dried – that is, transaction A is the same as transaction B and the same as the next day's transaction C – the company doesn't necessarily need to look at innovative solutions.

But when a company operates in a very dynamic environment, where rapid product changes and millions of transactions take place on a recurring basis, it is important to review updated solutions regularly.

The difference is similar to the difference between building a new railroad and keeping the trains running. If you are building a new railroad, you always look for better, faster solutions that are more robust and less costly. And they do exist.

You should strive for an absence of any compromise on your site – an interesting puzzle. One of the challenges in the security business, both in the physical world and in the cyber world, is how to measure a negative. How do we determine the number of burglaries prevented by having a police car drive up and down the street every so often? In the online IT world, having your site up and running for a long time with no breaches or compromises is a solid measure of successful security.

Preventing Fraud

An interesting difference exists between doing business in the online world and doing business in the physical world. Often in the physical world, the seller is forced to deal with fraud issues, as when a customer pays with counterfeit money or a fraudulent credit card. But the buyer generally goes to a known place. Online, sales depend on mutual trust, and often trust comes with the reputation of the seller, as well as the ability of the buyer to ensure the transaction is valid.

It is important to look for anomalies, to zero in on situations where people are identified as not providing what they agreed to or using a fraudulent credit card. Clearly, by sharing information among various business units and platforms and with the law enforcement community, we can reduce the amount of fraud.

How to Determine What Is Acceptable

On e-commerce sites, people are required to accept guidelines before they are permitted to participate in transactions. I think it is important to be able to review the history of those with whom you do business. We have all heard the adage about learning to walk before you run. If a new business opens in your neighborhood and the business is not a brand name, people tend to buy a little at a time to gain trust in the store. The branding of a business for being a trusted place is invaluable.

When you buy items at a swap meet or a store, the transaction is between you (the consumer) and the retailer. When you go to a mall and make a purchase, the mall itself is not responsible for the product you bought. In the future, there may be a need to facilitate disagreements, and dispute resolution may come through third parties and the credit card companies. Basically, it is a good way to do business. Clearly, though, the transaction results in a relationship between the buyer and seller.

A business must have a clearly articulated policy that covers dispute resolution, including when and where it takes place, who arbitrates, and how the problem is resolved.

The role of a CSO/CISO is multifaceted. I must look at the integrity of information technology systems, both internal- and external-facing. The internal IT organization, which employees depend on for their ability to do day-to-day business, often provides back-end access and control to the Internet-facing part of the business. The IT group makes sure that systems are engineered securely, that we have audit and compliance in day-to-day activities, and that we have appropriate policies so that even after we engineer something directly, it is maintained in a secure manner.

One Major Security Nightmare

For most of us who have been in the security business for a long time, one of the greatest security worries comes down to not knowing what will happen next. Where there is vulnerability in an operating system or a piece of hardware that we *don't* know about, someone else may be able to exploit it. That causes us to have to change some of the ways we do business.

We have an excellent record in dealing with problems we have seen in the past. What will happen in the future worries us most: We worry about the unknown and the unpredictable. The most common unknown seems to be how and where another vulnerability might pop up and how long it will be before the vulnerability is discovered and exploited.

Learning from the past gives us a way to prepare for the future. Taking a page from military strategy, where leaders look at wars and battles and conflicts of the past to determine future moves, we examine past security situations and try to build them into scenarios that may affect the way we do business in the future.

When the Melissa virus first hit, it took advantage of what was a great feature in word processors and turned that advantage into a negative result. You can make a similar problem less likely to happen in the future by changing the way you allow macros to run. When email attachments started to become malicious, we started blocking email attachments at the virus walls, or we put rules in place that didn't allow someone to just click on something, and we ended the evolution of that type of attack. We have to try to ensure that we don't make the same mistakes that were made in the past, which often results in changing some business processes.

Challenges in the Marketplace

There are a couple of challenges in the security marketplace. First, there is the ever-changing need to adapt your business environment to the threat environment. We know there are hackers; we know there are viruses and worms; we know there are denial-of-service attacks. Unfortunately, in the IT world, and particularly in the information security world, the threats are not always clear. In a conflict in the physical world, when you see troops building up on the border, you know something is about to happen. Or if someone launches a rocket, your radar can detect it, and you have some time to prepare.

In the IT world, we saw corresponding problems with the Sequel Slammer in January 2003, where it gained maximum penetration within ten minutes. One of the Achilles' heels of the online world is that it is so ubiquitous, and things happen so instantaneously, that situations turn malicious and spread with phenomenal speed.

How to Approach Risk

One type of risk is that of simple perception: that bad things *will* always happen. I have been buying products online for years and years and years. I have never had a bad experience either with a purchase or with compromised personal information or credit card information online.

In the physical world, I have had many bad experiences. For instance, I bought an earpiece for a mobile telephone at a mall one day. A couple of months later, I noticed that every few weeks, there was a charge for a few hundred dollars that I didn't make on my credit card. It turned out that the store employee who took my credit card was stealing cash and writing a receipt against my credit card.

Another offline problem occurred when I stopped at a restaurant in Texas a few years ago. Within days, I started getting charge bills from gas stations and restaurants and department stores in that area. Someone evidently took the credit card receipt, got the numbers, and started using the card.

The risk is that people believe that kind of theft happens all the time in the online world, but in reality, it happens only very rarely, especially compared to the physical world. As a result, we do not achieve the full benefit of the value of the online world.

You can envision the literally billions of dollars' worth of transactions that occur daily in e-commerce. There is fraudulent activity in only an extremely small portion of that business.

The second major risk comes from trying to push out products before they are ready for prime time. We have seen this happen in the IT industry for a long time; there is a driving desire to get new features built into something and get it out the door before some other company does. Fortunately, most IT vendors now make it a corporate policy that security comes before features.

Another risk is not realizing that the whole infrastructure is a part of the larger ecological system of e-commerce and the online world. Just as we have seen with power blackouts recently in the Northeast, an infrastructure problem can stop everything from fuel being pumped to sales at cash registers. That is an indirect effect of the IT community. Having a better understanding of the interdependencies helps mitigate some of that risk.

Strategies to Overcome Challenges

There are a number of strategies to beat those problems. One is to make security part of the responsibility of an executive leadership team and not bury it somewhere in an organization.

Another strategy is to provide security education and awareness to all – from the executive staff to the person who just comes in to do word processing on a computer system. Everyone has to understand that they are part of the security infrastructure and that they have to do their part to secure their piece of cyberspace.

An additional strategy is to make it a policy and a part of the process that security is built in at the very beginning and not added on later. Designing and engineering security into systems and processes from the outset is critical to success and provides a much better ROI.

Finally, in case the unthinkable happens, companies must have a response plan and a continuity plan in place to minimize the impact on the company's ability to use its IT systems.

How you spend your time in an organization dealing with security issues depends on where you are in the organization.

A security executive spends an inordinate amount of time dealing with education and evangelizing about security and why it is important in a particular business. A line manager's time is spent managing the ever-increasing requests for security resources. It is important to be proactive in explaining to people the benefits of engineering product security into new projects from the start. Once they get that concept and have many projects going, then you have to be there for every one of them with resources.

For the individuals who are actually doing the work – configuring firewalls, developing anti-virus solutions – security is a real challenge. As new threats come out, they have to stay abreast of what is going on.

Major Misconceptions About Security Issues

One of the biggest misconceptions about the business of security is that you can achieve 100 percent security. As the saying goes, "Security is not a destination, but a journey." It takes constant vigilance to ensure that security is maintained. Another misconception is that every time you go online, your security will be compromised.

In most instances, particularly for large, well run e-commerce operations, bad things very rarely happen. In an e-commerce portion of a brick-and-mortar type of industry or a company that is totally online and totally IT-based, most companies work diligently to minimize the impact that might occur from online transaction problems.

A serious misconception about security products and vendors is that there is a 100 percent technological fix to any security issue that currently exists. Security is not only about technology. The people-and-process part of it makes security policies work. People must be trained to maintain the systems, and others must be trained or educated to use the systems.

A company has to have policies and processes in place to make sure security is maintained. A former colleague of mine, Scott Culp, put together what I think is one of the most brilliant documents ever created about computer safety measures: "The Top Ten Immutable Laws Around Security."

One law says not using updated virus signatures and antivirus software is only marginally better than not having antivirus software at all. This law is critical: The misconception is that you have to install a security program only once and it is good forever. Again, security is not a destination – it is a journey. Like anything else, it requires maintenance from time to time. It's imperative to understand that principle.

Staying on Top of the Security Game

To keep up with the latest on security, it is important to maintain a close association with your peers in the industry. In this industry, CSOs all follow virtually the same daily routines. But our environments are different, and some of the technologies we deploy are different. So the ability to communicate and share information with our peers is vital.

For example, we recently created a Global Chief Security Council. The Council's ten charter members are Bill Boni of Motorola; Vint Cerf of MCI; Scott Charney of Microsoft; Dave Cullinane of Washington Mutual; Mary Ann Davidson of Oracle; Whitfield Diffie of Sun Microsystems; Steve Katz, formerly of Citigroup; Rhonda McLean of Bank of America; Will Pelgrin of the New York State Office of Cyber Space Security; and me. These security professionals have been in this business for a long time, and we are creating a think tank for all of us. We will share with each other not only the challenges we face, but also some of the benefits we see.

Our purpose is to mentor people coming into this relatively new career field, particularly at the executive level, where there are such jobs as chief security officer. Three years ago, only a few of us had that title or similar ones. We would like to make a contribution to the success of those in this new role, based on our successes or – just as important – based on failures we have seen.

Staying current in this field involves attending conferences and being aware of training and resources and development activities at universities. Some of the major universities have fantastic information security and information assurance programs. Staying in touch with them and serving on some of their advisory boards keep us in the loop and helps them develop better educational and R&D agendas. It is also important to work closely with non-traditional universities, like the University of Phoenix, which specializes in adult education that merges business acumen with security technology. Additionally, we work with various schools to develop their curricula for the next generation of security engineers.

Many of the professional organizations, such as ISSA (Information Systems Security Association), ISACA (Information Systems Audit and Control Association), and IIA (Institute of Internal Auditors), keep professionals informed in areas where they simply would not have the time to do research on their own. They help bring important topics to the forefront to keep people updated on security issues. The annual SANS Conferences on network security and working with security experts in government provide opportunities for exposure to a great deal of experience in cyber security.

Changes in Security

Fundamental changes in security in the recent past are affecting security today.

First, our connectivity is greater now than ever, but it is only a small portion of what it will be in the future. Consequently, a security event that used to cause some disruption locally now has the potential to cause a more widespread effect.

Second, our dependence on the day-to-day critical infrastructure is greater than ever. So there is a much greater need to ensure that security is ubiquitous from the design and engineering stage to the maintenance cycle, through the management cycle, to the end user's application of the IT system. Security has evolved from being recognized as a necessary evil to being a core business process. One of the key success factors is making security a routine part of daily operations.

Third is heightened security awareness. Security used to be something only security people worried about and were charged with. Now it is a regular boardroom discussion topic. It permeates organizations, and any successful company must have a "culture of security" that is as strong as the rest of its corporate culture.

Other changes will have profound effects on security matters.

The tools we currently use to monitor various aspects of security will be based more around the Web services model, where anyone can have security audits on demand. These will be much more valuable and meaningful than the traditional annual audits. We will become more adept at automating security and features into systems, so we don't have to spend time taking down systems and repairing and rebooting them.

A major issue revolves around authentication, as in smart cards and two-factor authentications, and moving away from an environment where static user IDs and passwords provide security. To be secure now, you have to have dozens of passwords for all the different systems you have access to because that is the right way to do things. Too often, we choose easy passwords to help us remember them. The changes in security will need to include the ability to do strong identity management, which will involve much more than just computer access. Our digital identities will be needed for many other routine functions.

One last major change is the ubiquitous nature of mobility – the whole wireless world we have moved into and receive great benefits from. It will get better, more secure, and more robust, so we will be able to do much more from anywhere in the world and do it securely.

A Vision of Security's Future

We need to see a more rapid deployment of Internet Protocol Version 6 and the security features within IPv6. It fixes a number of problems, such as the limited number of available addresses, adds improvements in areas like routing and network auto-configuration, and will allow our key Internet infrastructure elements – the top-level domain name servers and the border gateway protocol – to communicate. As more and more devices receive IP addresses, there will be a greater need for those devices to stay accessible. That will change the way our top-level domain name structure is secured, as well as the technology surrounding it.

The purpose in that change is to affect reliability issues. When I was growing up, the local phone book was an inch thick. Now it's four inches thick. It is a big, heavy thing so cumbersome it's difficult to use. As we have more IP-enabled devices serving us, they will become more difficult to use because our systems were not designed for this widespread user base. Revamping some of the new technologies relative to IPv6 will give us the ability to manage that more effectively. Better encryption, security, and authentication are the three components that will make up that new system.

A number of organizations, like the IPv6 Forum, have been trying to roll the snowball uphill for several years. They are finally reaching a point where we can see some advances. The U.S. Department of Defense, for example, will release some guidelines on the transition to IPv6. We are starting to see some forums in Europe and Asia that are serious about developing the next generation of IPv6-enabled hardware and software.

A number of the improvements I would like to see in the future are already being developed, and I am excited to see many of them moving toward completion. For my vision of security in the future, we need to have a clear R&D agenda that is jointly developed by the private and public sectors. They are making our lives more efficient and connected, and they become business enablers.

One example of the convergence is the PDA as an integrated-on-one device. I finally have reached the point where I have a device that combines mobile phone, email, calendar, and other features. But I still carry a cell phone and a PDA because the combination device doesn't work in all places where I travel. Getting integration and voice-over-IP finished and rolled out will make communication more ubiquitous, which will benefit millions of people.

Best Security Advice

I most often talk to people about the culture of security by using a seatbelt analogy that a friend of mine shared with me. At one time there were no seatbelts in cars, and when they were first introduced, people were hesitant to use them. Some didn't think they needed a seatbelt; some never thought they would be involved in an accident; some thought if they were in an accident, they would be trapped in the car, which might make their situation worse. And some just didn't think it was an important issue. When manufacturers put buzzers in the cars, we learned we could turn off the buzzers by clicking the belt behind our backs.

Eventually we became more educated and more aware about how seatbelts do save lives. We have adapted our culture to the point that just a few weeks ago, I was invited to give a speech at a university located where my son and my grandkids live. While there, I asked my grandson if he wanted to go in the car with me or ride in the car with his mom and dad. He said, "I'll go with you, Grandpa, but do you have a car seat?"

In our culture, seatbelts are now normal for us. What's the first thing a six-year-old child does upon entering a car? Buckle the seat belt. What is the second thing if you don't buckle yours? They tell you about it.

Golden Rules of the Business of Cyber-Security

First, security must be engineered in at the very beginning and has to be part of the core business process.

Second, security must be maintained by assessing the status of security daily.

Third, people must be educated continuously on what they can do to protect themselves.

Fourth, security must be part of the core business process by building it around crucial IT systems.

The way I like to describe this collection of rules is *The Mosaic of Security*.

There is a phenomenon in the art world where thousands of little photos are put into a big mosaic, so that when the viewer stands back, it looks like a picture of something big – a monument or a face. I see security as a similar process. There is a little photograph of a router, a little photograph of a person using a password, and many thousands more photos. All these different mosaic tiles are independent of each other and have their own qualities. When you put them together, you have a truly beautiful picture of a secure, trusted IT environment.

The Mosaic of Security draws on the strengths of the empowerment of each individual, business, university, and government to be one part of a bigger, more wonderful, and more secure cyber world.

Howard A. Schmidt joined eBay as vice president and chief information security officer in May of 2003. He retired from the federal government after 31 years of public service. He was appointed by President Bush as the vice chair of the President's Critical Infrastructure Protection Board and as the Special Advisor for Cyberspace Security for the White House in December 2001. He assumed the role as the chair in January 2003 until his retirement in May 2003.

Prior to the White House, Mr. Schmidt was chief security officer for Microsoft Corp., where his duties included CISO, CSO and forming and directing the Trustworthy Computing Security Strategies Group.

Before Microsoft, Mr. Schmidt was a supervisory special agent and director of the Air Force Office of Special Investigations (AFOSI), Computer Forensic Lab and Computer Crime and Information Warfare Division. While there, he established the first dedicated computer forensic lab in the government.

Before AFOSI, Mr. Schmidt was with the FBI at the National Drug Intelligence Center, where he headed the Computer Exploitation Team. He is recognized as one of the pioneers in the field of computer forensics and computer evidence collection. Before working at the FBI, Mr. Schmidt was a city police officer from 1983 to 1994 for the Chandler Police Department in Arizona.

Mr. Schmidt served with the U.S. Air Force in various roles from 1967 to 1983, both in active duty and in the civil service. He served in the Arizona Air National Guard from 1989 until 1998, when he transferred to the U.S. Army Reserves as a Special Agent, Criminal Investigation Division. He has testified as an expert witness in federal and military courts in the areas of computer crime, computer forensics, and Internet crime.

Mr. Schmidt has also served as the international president of the Information Systems Security Association (ISSA) and the Information Technology Information Sharing and Analysis Center (IT-ISAC). He is a former executive board member of the International Organization of Computer Evidence, and served as the co-chairman of the Federal Computer Investigations Committee. He is a member of the American Academy of Forensic Scientists. He serves as an advisory board member for the Technical Research Institute of the National White Collar Crime Center, and is a distinguished special lecturer at the University of New Haven, Connecticut, teaching a graduate certificate course in forensic computing.

He served as an augmented member to the President's Committee of Advisors on Science and Technology in the formation of an Institute for Information Infrastructure Protection. He has testified before congressional committees on computer security and cyber crime and has been instrumental in the creation of public and private partnerships and information-sharing initiatives.

Mr. Schmidt has been appointed to the Information Security Privacy Advisory Board (ISPAB) to advise the National Institute of Standards and Technology (NIST), the Secretary of Commerce, and the Director of the Office of Management and Budget on information security and privacy issues pertaining to federal government information systems, including thorough review of proposed standards and guidelines developed by NIST.

Mr. Schmidt holds a bachelor's degree in business administration (BSBA) and a master's degree in organizational management (MAOM) from the University of Phoenix. He also holds an honorary doctorate in humane letters.

Dedication: To my wife and best friend, Raemarie, who is truly one of the leaders in the field of computer forensics and has given so much to train law enforcement to help them combat cyber crime. She has been my greatest inspiration, supporting my work despite much sacrifice on her part.

Protection of the Intangible:
The 21st Century Security Paradigm

William Boni
Motorola
Vice President & Chief Information Security Officer

The Role of the Chief Information Security Officer

Most large organizations survived for decades without a chief information security officer. However, by the end of 1999 it was clear to executive management at Motorola that additional leadership was needed for what had become a crucial role. This change developed from the recognition that a new, more tightly integrated global program was required to cost-effectively manage the vastly increased array of risks that could affect the company's business.

As the first CISO, I joined the company in the spring of 2000, reporting to the global CIO, who reports to the president of Motorola. The role carries with it global responsibility for the development of security policies, process development, and design of the security architecture to safeguard the company's network operations, information systems, and digital forms of intellectual property.

To accomplish this, existing IT security engineers from the various business units and the small corporate policy team were gradually aggregated into the Motorola Information Protections Services (MIPS) department. This original staff of about 12 people has been augmented over the past three years with additional staff to bring the current total to about three dozen. Located primarily in North America, Asia, and Europe, in descending order of priority, the staff is divided roughly equally between direct support to major business units and enterprise security responsibilities.

One of the benefits of the consolidation of all security professionals into a single organization was the ability to capture and manage spending for security in one department. Additionally, it allowed for career development and rotational assignments for the security staff. The security team is organized along functional lines into protection/prevention, detection/response, program management, and privacy. Although the total budget is approximately 1 percent of IT spending, it represents about 75 percent of the company's total spending on information security. Additional staff in the IT operations group is responsible for the configuration of firewalls, Virtual Private Networks (VPNs), remote access security systems, and other infrastructure services.

The protection organization, in addition to the above, also acts as advocate for essential security and privacy safeguards in global business operations as well as product and services development. The staff engineers in the protection team work closely with product and service development teams to advise on issues and solutions that can be integrated into the company's new and existing offerings. The research and development staffs are well equipped to design products that address traditional customer requirements. The security organization brings to the mix an intimate understanding of the risk management challenges facing a large, complex global enterprise. This allows us to function as a surrogate for a typical enterprise organization. In that capacity we frequently work to alpha test new devices in our production environment and provide feedback and recommendations to the product development teams.

A second critical role of the protection organization is to ensure the company has adequate programs to deal with the risk of catastrophic failures that could affect the continuous availability of critical information and systems. This is a significant step beyond the conventional disaster recovery programs that were well developed in the last decades of the 20th century. The difference is that disaster recovery for IT tended to be "application" centric. A respectable process was to focus on fully developed disaster plans for only the most critical applications, while others, perhaps with less stringent recovery time objectives, would rely on offsite media and rudimentary recovery procedures that would be supplemented by the institutional knowledge of the support staff when necessary.

This meant that many plans rarely considered the potential for wholesale loss of complete data centers and their staff, or the impact of prolonged regional events spread over multiple states or even the entire United States. However, as we learned to our horror on September 11, 2001, even manufacturing organizations must now be prepared for catastrophic loss of entire facilities and of much – perhaps even all – of the staff "institutional knowledge". Driving the lesson home, the regional power outage of August 2003 proved multi-state, regional disruptions to the critical infrastructures in the United States were possible.

The protection organization now works with both business continuity planning and crisis management teams to create integrated planning and testing protocols that will help manage the potential for and impact of catastrophic events on the company. The tight linkage of IT disaster plans with these other dimensions or risks are improving the company's overall protection profile and reducing the financial and operational consequences of events that may never manifest themselves, but if they do, could be extremely harmful.

Recently the security team has been tasked by the CIO with leading the company's overall program to define and implement a comprehensive privacy and data protection program. Although privacy and information security might seem a strange combination, we have found the privacy safeguards complement the established efforts to safeguard proprietary information. Information considered sensitive because it is personal must be protected consistently with the policy standards of the company.

Our assessment is that much of the risk to the company for privacy issues arises from the operation of electronic business sites and a selection of key applications relating to internal staff and external consumers. Our approach has been to work as part of a multi-disciplinary team including human resources and legal professionals to develop a policy and appropriate processes and procedures to ensure compliance with the policy in both technology and operations. We have found that beyond policy, many of the privacy assurance mechanisms are in fact applications of security technology, especially access controls and audit trails.

As the above discussion about catastrophes, business impacts, and privacy reveals, the protection organization increasingly functions as "risk management" advisers to executive management for all risks, threats, and vulnerabilities related to technology. We believe this is a "mega trend" that will increasingly affect existing and future security and protection organizations.

Another major trend that has had an impact on our security program and will become critically important to other organizations is the dramatic increase in the use of outsourcing as a specific tool to support the company's strategic and tactical objectives. As a consequence of the increased role of outsourcing over the past three years, the charter and role of the security team have changed substantially. Beginning in early 2000 and coming to the fore in the last year, the company has completely revamped the business model. In the past it operated closed internal business processes where all major business operations were performed by company regular staff.

However, as the intensity of global competition increased, the company found that it gained significant competitive advantages through the flexible and adaptable capabilities available through outsourcing and was better able to deal with rapid and frequent changes. The company has formally adopted a global sourcing model for IT and business operations to leverage global talent diversity and cost structures to best accommodate customer needs. This resulted in major restructuring and realignments of business operations.

Given that IT operations were not considered to be a core competency of the company, in the spring of 2003 the company initiated a ten-year outsource contract with one of the leading U.S.-based providers of IT infrastructure services. In tandem with this effort, the company has increased use of offshore IT development organizations in low-cost countries, while still retaining key elements of IT to support product development. To provide direction, oversight, and governance of this multi-sourced model, the company created an IT leadership team focused on strategy, architecture, program management, and security.

The impact of these changes on the role of the security organization has been profound. We have focused significant efforts on creating a comprehensive governance framework for the overall information security program to ensure compliance with the company's security policies and procedures in the multi-sourced model. We are quickly moving away from a solutions and project perspective and emphasizing much more the definition of and compliance to requirements and company policy standards.

These changes have not diminished the need for security and privacy professionals; on the contrary, they have increased demand for expert advice and oversight. They have, however, shifted the desired skill sets for security professionals to include a well-balanced complement of technology expertise, process skills, and business acumen. The ideal security professional is now perceived to be a managerial staff engineer, well grounded in one or more foundational security technologies, with equal parts technical skill, business savvy, and appreciation for legal and regulatory challenges in the global marketplace.

As a consequence, the protection organization has begun divesting itself of operational responsibilities. As these are moved to the outsourced IT organization, we are focused on re-tooling the retained staff with the education, skills, and knowledge required to succeed in the new business model. This means a combination of training in new technologies and increased emphasis on building business knowledge that allows better risk decisions.

Governance of IT Security

The change in the responsibilities of our information protection professionals has driven in changes to the overall governance structure. We believe the following model represents a good baseline for effective information protection governance. As more organizations grapple with the impacts of outsourcing, partnering and global competition elements of this model can be adapted to provide effective management control over the security status of the entire "virtual" organization.

On a regular (at least annual) basis the chief security officer or chief information security officer (whoever has primary accountability for safeguarding the enterprise against information-related risks) should be obligated to report to the board of directors or the audit committee of the company on the overall status of key risks and safeguards. Effective governance starts with having that kind of accountability imposed upon the executive leadership of the organization by their external oversight authorities. This is important to show that the board of directors has exercised its responsibility to ensure that the company's management, operations, and practices are consistent with the shareholders' expectations and interests.

The same set of issues that resulted in the Sarbanes-Oxley legislation in the United States has increased the attention to related governance topics, such as information protection. There is renewed attention to ensuring that governance is effectively exercised by all of the appropriate levels of management.

However, in every organization there is a struggle to obtain sufficient resources for risk management at a time of economic downturns. The most recent downturn in the U.S. economy that began with the collapse of the dot- com era in early 2000 occurred at the precise time that risks to company systems reached unprecedented levels. One huge challenge for the CISO is to find ways to manage the increasing range of risks without expecting or obtaining huge increases in budget and staffing. The organization should spend what is required to manage risks that have significant impact, but no more.

A key responsibility for every CISO is to make sure executive management understands which risks are in-scope and therefore addressed by the protection program at its current level of resources, and those risks that cannot be effectively managed and are therefore accepted by business management. The best way we have found to do this is to create a risk management matrix that graphically represents the highest-impact risks that are managed, as well as other important risks that are not being managed. The prioritized list of risks that remain to be managed are tracked to the organization's roadmap for the next two years of security and protection investment and may be re-introduced to executive management as "business case based spending" if circumstances change and require immediate protection efforts.

Since the company now operates in a multi-source IT environment, we have developed an approach to ensure we instantiate a common global approach to managing risks through the application of our baseline security policy framework. Although we have outsourced key elements of our IT infrastructure and leverage offshore development capabilities, we have also retained IT staff that provides specialized capabilities, depending on the business unit. Ensuring that all of these elements comply with the security policies and standards is part of the CISO responsibility and represents a very dynamic, challenging area. The following illustration is a graphical depiction of the IT Security Governance Model that has been developed.

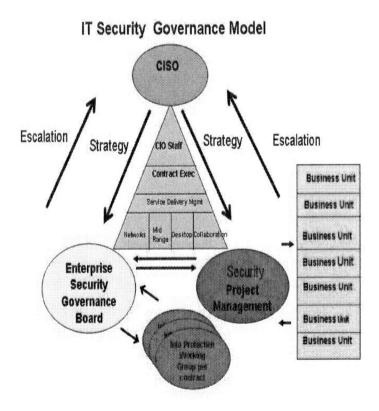

Protection Technology

The majority of the protection products that are installed are off-the-shelf products. We expect to continue that into the future. There are some products that we have helped co-develop with companies in which Motorola has taken an equity position. These provide our enterprise security console for correlation analysis and vulnerability assessment. We use a combination of off-the-shelf anti-virus software, network scanning, vulnerability assessment tools, system- and host-based intrusion detection software, and a specialized cryptographic application for securing electronic mail to help manage a wide range of risks to systems and the network.

From my training as a counter-intelligence officer in the U.S. military, I learned that a great deal of useful information is available from the open or public media. These sources can help identify the nature and extent of threats to organizations. After years as a security practitioner, one lesson has become very clear: If company executives are reading about a security problem or a cyber incident in the *New York Times* or the *Wall Street Journal*, the security organization had best be prepared to deal with the impact of that same problem on their organizations.

There are so many theoretical threats, risks, vulnerabilities, and dangers that a security officer could go mad trying to ensure that every potential avenue of attack is secured. In fact, that is rarely necessary for most organizations. Outside of financial services, and more recently health care providers, there is not yet a commonly agreed-on baseline set of security standards in the United States that applies to all business organizations. The standard of due care will most likely derive from what is commonly known and applied by the practitioners in the field, which is in large part information that is available via commercial and trade press resources.

One of the first things a security practitioner must do is to stay current with what is publicly available concerning how organizations are actually being attacked or victimized. Such information may be found in the business press, as well as the technology and trade media. There is now a wealth of additional and relevant information available for free or for a modest price through online services and specialized intelligence and alerting services. Together these resources can provide valuable sense of how risks and issues are developing and suggest where the protection officer ought to focus attention next.

Another part of the job, which has become increasingly important, is to stay actively engaged with the governmentally sponsored sources of vulnerability and threat information, such as Information Sharing and Analysis Centers (ISACs). Although official sources have not replaced and may never be able to replace public information, as homeland security becomes more deeply rooted in the United States, they can become an effective forum for reporting suspicious activity or unusual events.

One of the most useful sources of information to senior security practitioners is their security colleagues and counterparts at similar organizations in their industries. For a long time, the underground has had an advantage over the "white hat" community simply because they use the medium of the Internet to exchange information relevant to their perspective, which is how when and where to attack.

It has only been in the last three to five years that the security community has finally begun to realize they can share information to common advantage. Even when business organizations may be competing, they can share a common cause against those who would vandalize or disrupt their electronic infrastructures. Some of the best ways to share this kind of information include active involvement with the FBI's InfraGuard, ISAC membership, and attending and presenting at security seminars. The threats are global, not simply a domestic issue. We have created a global infrastructure called the Internet that allows threats, vulnerabilities, and protective information to flow rapidly to those who are actively engaged with the potential source of information.

Assessing Risk

There are many kinds of risks that must be managed. To make the tradeoffs in various elements of risk, we look at security holistically. We use the classic frameworks – basic principles, such as confidentiality, integrity, and availability are considered baseline controls. Part of the CISO role is ensuring that appropriate product and service solutions address the key elements that can put the business strategy or business operation of the enterprise at risk. Our strategy is summarized as executing policies, products, and procedures that "prevent, detect, and respond" to threats that can affect the enterprise. What we are looking for is: How do we manage risks to the most important sensitive, proprietary trade secret and digital IP (intellectual property) base assets of the company? How do we prevent threats to the operational availability of those assets?

Basically, we build a matrix with prevent, protect, and respond. Within each area, we are looking for solutions that help us manage risk and the most significant threats to those particular interfaces between those axes of the matrix. You put prevent, protect, respond down the vertical axis and put C, I, and A across the top and figure out how you prevent threats to the confidentiality, the integrity, and the availability. You figure out how you can detect risks or threats to all of these and how you can provide responses to all of these factors.

	Protect	Detect	Respond
Confidentiality	Encryption	Vulnerability Scanning Tools	Computer Forensics
Integrity	Anti-Virus System	Correlation Console	Computer Incident Response Team
Availability	Backup Systems	Network Monitoring	Disaster Plans

A company that moves fast, with many business units, operations, and product lines, faces the challenge of not being able to manage all risks through all lines of business and all operations equally. Given that we follow the strategic imperative stated by Frederick the Great, which is, "He who defends all defends nothing," rather than spreading ourselves too thin, we focus where we can get high leverage and high payoff. We are always looking to cover 80 percent of the risk and accept the last 20 percent, rather than try to drive risk completely out of the business. If that is the goal, the organization is at risk of introducing potential losses. The company also reduces its ability to adapt and respond rapidly to marketplace and business changes.

We believe security is inherent in the brand promise of our product lines. The heritage of the company began with the production of radio systems for Illinois police agencies in the early 20th century and continued with major contributions during World War II with the development of the handy-talkies and so forth. The legacy associated with Motorola as a brand implies that security is offered by the products we manufacture. Therefore, ensuring that we have responsible levels of protection for the functions provided by the product is an important means of enabling the revenue of the company.

Companies whose products are repeatedly vandalized or subjected to high levels of abuse or exploitation by criminals or others will likely find their sales suffering. In our case, since we are involved extensively with cellular communications and broadband digital communications markets, to the extent that we are able to provide secure mechanisms to the operators of those communications infrastructures, they are able to offer customers additional products or services that can generate incremental revenues. So the customer enjoys greater diversity of choice, and the operator enjoys the opportunity for increased revenue, enabled by a secure channel for distribution of additional content or service.

Selecting Security Products

In selecting security products, the challenge is in balancing attention to innovation, which most frequently comes first out of the startup and venture-funded entities, with the stability and longevity more commonly associated with larger, better funded, more established organizations. Typically it seems that the best innovations come from startups, but the concern is that they may not survive for long, and if they do survive, they are most likely to be acquired by one of the bigger market players, anyway.

Typically we evaluate risks for our company annually. We identify, using available information, the priorities for which we will acquire solutions. When we agree on that, we conduct product reviews with a reputable advisory service.

From this preliminary assessment we identify which solution or solutions look like they address the broadest range of our requirements, based on a preliminary determination by the security engineers and the IT security architecture team. To be considered, products and solutions generally must have earned a favorable review by one or more of the leading analysts. We rely heavily on the analysts to filter the seemingly endless list of potential products that could be either promising solutions or dead ends. We simply are not interested in being the beta tester of a next-generation product, as we do not have time to be a laboratory to trouble-shoot someone else's product.

When we have a candidate solution, it is presented to the internal architecture and strategy review board to validate that it will align to our overall corporate operating environment and roadmap. When that approval is obtained, an initial pilot is conducted in which our goal is to determine whether the solution performs as described. The criteria we evaluate are fundamentals, such whether the product actually does what it claims, whether it performs in a consistent, reliable, repeatable manner, and most importantly, whether it scales to our environment. Very frequently solutions are developed that are excellent for small and medium organizations, but Motorola is very large. We are in 64 countries and have nearly 100,000 employees, more than 10 million IP addresses in our network, and nearly a quarter-million devices on our network. If we are looking at a solution that will interact with a broad cross-section of that environment, it has to be scalable, and it has to be able to operate in a decentralized, follow-the-sun mode of operation. Since our security organization focuses on managing risk at the global enterprise level, the solution has to allow us to pull the data into a centralized location where we can analyze, react, and respond to it on a rapid, timely basis.

We look for solutions that have a high probability of being successful in our environment. We have a very limited staff and budget and a huge array of risks to be managed. We cannot spend a lot of time trying to tinker with a product to see if it performs. If it passes basic functionality, then, given our unique operating environment, we determine how it scales and interacts with the other applications and components of our global environment.

We tend to categorize products and solutions either as prevention, protection, or response technologies, using the protection matrix as noted above. Prevention technologies harden the host/network or application against exploitation. I tend to put things like vulnerability assessment technologies there, as well as sensor technologies. So the vulnerability assessment technologies go out and probe the individual platforms that are vulnerable to exploits. Part of the challenge is that when you take a look at the huge global network and operation around the world, scanning that type of environment with a particular scanner itself produces millions of items of vulnerability information. That has to be quickly triaged down to the ones that have the most significant risk so you can put your remediation methods against those elements. Things like disaster recovery planning tools, business continuity planning tools, and the ability to restore systems that have been damaged or destroyed allow you to investigate incidents so you can have a better probability of determining the root cause or accountable parties for a particular incident or event as it occurs in your environment.

Important Issues and Challenges

At the beginning of the 21st century, we are in a unique period in history. The vast majority of senior management in global organizations earned their promotions to the executive ranks in the

20th century. Inherent in that 20th-century experience is a certain tolerance for risk based on the events that are culturally common knowledge or common, shared experiences of society during the 20th century.

Wars were fought between armies; computers were introduced late in the 20th century; and if they didn't work, you found alternate ways of getting your work done, and that was acceptable. Tangible, physical items were both a source of wealth and economic prosperity or threat and damage. A tank was a vehicle for destruction; an airplane was an engine for either commerce or conflict; but both were quite tangible.

We have now crossed the threshold into an era in which the intangible – the electronic and digital – are the basis of commerce and the key enablers of global communications. But I believe they, like the airplane, will also become key elements in conflicts in this century. However, with a basis of experience grounded in material, many management teams don't yet fully comprehend the implications of this change in the gut, where it really matters. They have not yet internalized these changes. We are just now getting to the point where people are starting to realize that things such as computer viruses and worms actually are serious problems – but take a look at the speed of propagation.

I was doing computer security work when the Morris worm was released. That was the first example of propagation of a threat via network connectivity. A few thousand systems were affected, and the community was up in arms because of this. I attended early lectures by Dr. Fred Cohen, who is known as the "Father of the Computer Virus" and who wrote his doctoral thesis at the University of California. In the early 1980s he proved that self-replicating code could potentially be used as a weapon of conflict between nation states and military forces. We are now in a time when these threats are starting to actualize. We have experienced the "Code Red," "SQL Slammer", and "Blaster" – *and more are certain to follow.*

Plotting the speed of propagation of computer worms and viruses shows that we have gone from a situation in which it took weeks to days, days to hours, and then hours to minutes for a threatening piece of malicious software to spread throughout the global Internet. Managers in many companies still don't fully appreciate how rapidly these attacks can spread and how potentially disruptive and destructive they may become. As a society we may have been led into a false sense of security because we haven't seen airliners falling out of the sky and trains running into each other as a consequence of cyber attacks – not yet, at least.

Many business managers think that "digital dependency" isn't truly dangerous, that the gains in productivity and effectiveness carry only nominal risks. My biggest concern is that the often unconscious acceptance of unknown risks is becoming more dangerous to the global society, as every day we become a bit more dependent on the Internet and networks of computer technology. Most organizations depend heavily on off-the-shelf operating systems and applications. As there is an increasing prevalence of common platforms, networking protocols, and applications, an exploit that is developed against any part of that infrastructure may be propagated throughout the Internet in minutes.

The fundamental underpinnings of the global applications and network communications that exist were not designed with high degrees of security or assurance as a fundamental requirement. So we are now in the process of trying to retrofit at least a basic level of protection into all of the components of the infrastructure. We are far behind in addressing the risks, which are simultaneously increasing. Perhaps over the next 20 years, as key vendors take the issue more seriously, and as the Department of Defense and U.S. government encourage increased security capability in the products they use, the situation will improve.

Unfortunately, many of our protective countermeasures have been developed based on a largely manual response protocol: (1) There is a problem; (2) we become aware of it; (3) we contact our vendors; (4) they provide a solution for it; (5) we deploy the solution; (7) repeat step one. Under the best of circumstances, with just a minor virus or worm, that can take 24 hours. When it took months or years before an attack happened, that was an entirely acceptable time for response. However, now the situation is that attacks are coming quickly after vulnerabilities are discovered and spreading in only minutes. In the absence of better automation for vulnerability remediation or mitigation, the advantage will remain with the attacker or intruder. Even where automated tools exist, there may be areas in the organization where updated protections cannot be pushed through the infrastructure. At the present it is often necessary to have support personnel go in and reboot systems, which may further disrupt operations.

Some security experts rile against organizations that have failed to patch their systems against the latest vulnerability and exploit. For many organizations, patching has typically and rightly been at the bottom of the priority list. Getting new services online, getting new locations connected and maintaining connectivity between operation components, and running a global 24x7x365 organization tend to take higher precedence, and appropriately so. On average, if an organization receives one major patch every week, for every major operating system or platform that is operated, it would take thousands of hours of a dedicated IT staff to patching to make sure everything is fixed within 24 hours.

So the advent of enhanced prevention technologies or ways to create what amount to virtual fire breaks quickly in the course of an incident, and doing that with computer-based intelligence rather than human-based intelligence, is one avenue where protective technologies must be improved. Having a human in the loop is a good thing unless it prevents a timely response to an automated attack. If we ever have a hyper-speed worm attack that actually contains destructive payloads, the result could be horrific unless such defenses exist.

Those risks trouble many security professionals. There is an imbalance between the protective capacity deployed in most organizations and the potential threat. The actualized threat has not yet demonstrated what may yet occur. When something like this happens, it could be extremely destructive. The security vendor community has become increasingly sensitive to these tradeoffs and the consequences. However, there has to be more demand from organizations – the businesses and academic and government entities – for improved levels of assurance in the products and capabilities they operate in their networks. When demand is well established, the market itself will help ensure that development resources are prioritized to deliver that kind of capability as a consistent part of the products and services delivered by organizations.

Even when an organization's management understands it must take additional steps to protect itself, the process can be problematic. The security solution sales cycle is generally long and typically requires working through multiple levels of management. From my experience, security problems and solutions are generally not yet very high in management priority, unless there has been a triggering incident. Even so, it will often require more than a year to complete an acquisition. Also, many organizations have multiple business units, each with its own risk management priorities. Even if they have identified a common security problem, sometimes one will prefer solution A, and another will prefer solution B. The challenge for the CISO is establishing processes to ensure that enterprise risks are managed and that legitimate and unique business unit priorities receive appropriate attention.

Future

A major challenge for the future is how best to align security priorities to make the greatest contribution to the business strategy of the enterprise. In the past, security organizations were frequently buried somewhere deep in the IT operations element of the company. They got their orders from the people who ran the data centers applications and networks. The ability to think and act strategically is vastly improved when the CISO sits on the same level as senior information technology officers from the major business units. This ensures that risks are given appropriate attention and folded into the business's overall IT strategy.

Affording the security role such access permits the security staff to more fully understand the business divers and focus on the priorities that will best support the overall success of the company. They can then consider which risks could jeopardize the strategic success. For example, would the premature disclosure of key components of intellectual property be crucial? Is continuous availability of certain information systems required? How might secure communications be handled between the internal and external members of the product design team?

Managing risks to business requires the security organization to monitor new and promising technologies that will be introduced to the company. For example, this year a key element of our IT strategy is to become an early adopter of Web services as a capability to deliver both internal and external services to Motorola. We are providing the framework for secure development and operation of Web services so that will facilitate the rapid adoption and effective utilization of the new technology. In this we help the business take advantage of the promise that new technologies provide while minimizing the potential risks. Providing policies, standards, and procedures for secure deployment of new technologies reduces risks to the company that could develop if employees were left to their own individual understanding about effective security protocols for Web services.

Overall public awareness of the importance of information security has increased significantly since the spring of 2000. The denial of service attacks against Yahoo! and eBay early in the year created quite a public stir. This was followed by the dot-com bubble bursting, which caused many people to downplay the significance of electronic business. It seemed that "electronic business" was not destined to change everything – at least not in the short term.

The impact of the terrorist attack on September 11, 2001, and the SARS epidemic in 2002 showed that it is important to have security and risk management professionals involved at the senior level of decision-making in the company. The understanding of risks and how to prevent or mitigate their impacts can help the organization deal with the unpredictable nature of the global situation. The security and risk management organization probably can't anticipate every scenario that could develop or threaten business operations, but they can develop an adaptive framework of best practices that includes capabilities in technology, tools, people, and processes. Under the right leadership they can then respond more effectively to any unanticipated event. When the organization becomes more adaptive, it should prevail more effectively in global competition.

There is a strong trend toward conversions of what might be classically described as physical security or corporate security functionality, which often includes facility protection, executive protection, and investigative functions. Those are combining with information protection aspects, which typically are IT-based, and an integrated business protection organization (BPO) is emerging. The potential convergence between the two security domains is obvious. There is real synergy that can develop if those organizational entities are combined. The vast majority of physical security practitioners have come out of a law enforcement or military background into those roles. The majority of information protection professionals have come out of technology backgrounds, and the two groups tend to speak different languages. There is a bit of oil and water mixing that can be very challenging. However, the organizations that make the effort and have the right leadership in place will find that the synergy between the two can create a framework that is actually far stronger and more capable of dealing with the compound risks of the 21st century.

Advice

One essential element of a successful career in security is to learn constantly. Try to think creatively in deciding how best to deal with a threat or risk. What appears to be obvious often isn't the entire story, and what appears to be trivial sometimes is a key to solving the problem. Also, take input from a variety of sources and expand it into an organized approach that will give good business value for the organization.

It is critically important for security professionals to think and articulate issues in business terms. This is a crucial skill because so many security practitioners came up through the technology ranks.

Their normal approach is to frame both issues and solutions in technical terms. Very frequently, whenever there is an issue or a challenge, they revert to technological language and technological solutions. But what is the business impact of the issue and the proposed solution? What does this really mean? Much of it comes from describing in English and explaining why it is important to the businessperson.

The guiding principle for success as a security executive is to manage risk. We use business impact as a measure of success. The hardest lesson for many information security professionals to accept is that it simply isn't their information. They are so passionate about it. Perhaps it comes from the technical background.

Our job is to make sure that risk is managed to a degree acceptable to the appropriate level of the organization's management. It does not mean that it has to be eliminated. Management gets paid to make decisions with their eyes open; however, it is the security professionals' job to *make sure* they do it with their eyes open! It's also important that they understand the potential consequences of not putting the recommended measures in place. At the everyday business level it is all about risk. It is not possible to eliminate risk from a business. However, a good security organization is like the brakes on a car: They allow the vehicle to drive faster and still arrive safely, to accept but also manage more risks!

William C. Boni has spent his entire professional career as an information protection specialist and has assisted major organizations in both the public and private sectors. For more than 25 years he has helped a variety of organizations design and implement cost-effective programs to protect both tangible and intangible assets. In a wide range of assignments Mr. Boni has assisted clients in safeguarding their digital assets, especially their key intellectual property, against the many threats arising from the global Internet. In addition, he has pioneered the innovative application of emerging technologies, including computer forensics, intrusion detection, and others, to deal with incidents directed against electronic business systems.

Mr. Boni has served as a consultant in several professional service organizations and now works as the vice president and chief information security officer of Motorola Information Protection Services. He is responsible for the company's overall program to protect critical digital proprietary information, intellectual property, and trade secrets. He also directs the people, processes, and technology programs that safeguard the company's global network, computer systems, and electronic business initiatives.

Mr. Boni has been quoted in leading print publications, such as the Wall Street Journal, US News & World Report, the Financial Times, LA Times, and CIO Magazine. He has also appeared on many network broadcasts, including Prime Time Live, CNN, and CNN/fn, discussing espionage and cyber crimes directed against American high-technology corporations.

Other assignments in Mr. Boni's distinguished career include work as a U.S. Army counter-intelligence officer, federal agent and investigator, investigator and security consultant, vice president of information security for First Interstate Bank, and project security officer for "Star Wars" programs and other defense work with Hughes Aircraft Company and Rockwell.

Mr. Boni is a past chairman of the American Society of Industrial Security (ASIS) Council on Safeguarding Proprietary Information, co-authoring both the 1999 and the 2001 ASIS-PwC report on "Loss of Proprietary Information." He is on the board of directors of the

International Society for Policing Cyberspace and served as an industry delegate to the National Cyber-crime Training Partnership (NCTP) and to the G-8 Cyber-crime Task Force, both sponsored by the U.S. Department of Justice. He is a member of the board for the Certified Information Security Manager's certification sponsored by ISACA.

In July 2003 Mr. Boni received a CSO "Compass Award" from the publishers of CSO magazine, recognizing him as one of a handful of thought leaders who "helped build security culture not just in their own organization but in the broader business community and the nation." He has also recently been appointed to the 2003/2004 board of the Computing Technology Industry Association. He holds the Certified Information Systems Auditor (CISA) from ISACA and the Certified Protection Professional (CPP) from ASIS International.

Mr. Boni is co-author of I-Way Robbery: Crime on the Internet and High Tech Investigator's Handbook: Working in the Global Environment, published by Butterworth-Heinemann in 1999. His most recent book, Netspionage: The Global Threat to Information, was released in September 2000.

Security for a Connected World:
It's All About Trust

Margaret E. Grayson
V-ONE Corporation
President & CEO

Security Is Fundamental

The security industry deals with people and their fundamental needs for safety and privacy. If people are going to work effectively in a technology-connected world and make full use of the communications capabilities available to them, they have to have some level of comfort and assurance that the private information they share remains private.

The security industry is about enabling people to communicate, using technology products and services that provide the necessary protections.

Securing Cyberspace Requires a Global Effort

The security industry is still in a developmental stage. The industry will evolve to meet the challenges of providing products and services that afford protection and privacy to its citizens. This end state will address both physical and cyber security needs and seamlessly integrate the public sector, private industry, and individuals.

The National Strategy to Secure Cyber Space, published by the U.S. government, addresses both data security and infrastructure security as a public/private partnership. This cooperation is an important component of the National Strategy for Homeland Security and is complemented by the National Strategy for the Physical Protection of Critical Infrastructures and Key Assets. The strategy highlights the need to engage and empower the federal government, state and local governments, the private sector, and the American people to secure the portions of cyberspace that they own, operate, or control, or with which they interact.

I agree strongly with that position and value the importance of the U. S. government in taking this step. I know, as a security professional, that cyberspace has no national boundaries. This is a global challenge, and will have a global solution. Through the responsible sharing of cyberspace by the world's citizens, all users gain the ability to work within an infrastructure that combines policy and practice in a form where trusted communication technologies augment personal privacy requirements. Ensuring the protections necessary to maintain the integrity of these vital systems is absolutely critical. In this early-stage industry we cybercitizens have to define what protections will be required as a basic entitlement and who will be responsible for providing them – the governments, an industry association or governance body, the community of security users, or some combination. The requirements will resolve into a combination of products and services that provide the protections and security to allow people to use information technology in their daily lives, conducting their business and communicating with others without concern that their private communications can be compromised.

Securing cyberspace is a difficult strategic challenge that requires coordinated and focused effort from our entire global society. In a world where continuous technology development is a given, educating users on the appropriate policies and procedures for risk mitigation and commitment by each user to maintain the needed education and knowledge is the key to realizing the full potential of *Security for a Connected World®*. (V-ONE Corporation)

Knowing When to Use Security Solutions

Looking at security issues from the chair of a CEO, as soon as you have a machine that will be connected to the Internet, or connected in any way, even internally, there is a potential need for information-sharing security. The question, "Is a security product required?" needs to be raised; the connection needs to be assessed; and the appropriate level of security needs to be in place to allow access to be managed and controlled by the information owner.

Security for internal usage considers the questions of where the data is controlled and who is authorized to access it. For example, in most environments the accounting systems are separate from operational systems, which in turn are also separate from an engineering or design database. You must assess very carefully the environment and what kind of access permissions or restrictions need to be imposed on that environment to make it function securely.

Cyber vulnerabilities are real, and they are everywhere. Anyone who has a network or even a single machine that is open to the Internet runs the risk of exposure to viruses and cyber attacks. Through the last ten to 15 years, the Internet has moved from the collegial environment that was envisioned when scientists and universities first began working with the Internet to something that is now available everywhere and used by almost everyone at some level – by governments, industry, and individual citizens. The Internet is not secure by its very nature. Recognition of the vulnerabilities is one of the first steps in the evolution of a secure cyberspace. For the Internet to be a useful tool for private communications, it needs to be secure.

When a network is open to the Internet, the first level of protection is often a firewall. In most situations, these networks remain vulnerable if external communication is important to the operating environment, even if powerful firewalls around them protect the internal assets. Intrusion detection systems can be added to let the organization know if someone tries to break in. Forensic network tools can be used to test the network for holes and vulnerabilities, allowing more safeguards to be put in place.

The Internet is a hostile place, and people have to be aware that it is hostile. But at some point you will have to open your firewall. You will have an entryway – a port open on your firewall to allow data to flow into and out of the network. Even if it is open only for email, this entryway must be open, yet secured. When you are sending email, you are sending it the way people would send a postcard, without even the protections offered by personally dropping it into the postbox. Anyone who wants to can read what is written on it or possibly even change it or destroy it. They can see the addresses of the sender and the recipient. If you want the message protected, you must seal it in a cyber envelope. You have to seal it with encryption, and you have to put a security stamp on it that sends it only to those who are authorized to see what is inside.

This is why organizations that enjoy the communication benefits of the Internet must go beyond firewalls and intrusion detection systems to anticipate and protect against potential external and internal threats before, while, and after the data is accessed. Awareness of the vulnerabilities of the Internet by its users is driving the security industry to deliver technology that enables the Internet to operate as a secure virtual private network (VPN).

Customers Need VPNs to Share Information

V-ONE, through its customer and industry interactions, understands that responsibility for data protection can conflict with the drive to share information, whether with employees on the road or working from home or with partners whose network configurations are unknown and non-accessible. The risks of opening the firewall are weighed against the desire to gain the efficiencies and cost effectiveness of the Internet for their business or organization.

Today's security products, especially the cyber security component, are evolving as people become more aware of what breaches of their privacy and security might result from an unwanted intruder compromising their private information or gaining unwanted access into their network. As a company that has been in the industry developing security software for more than ten years, V-ONE Corporation has experienced firsthand the evolution from the very early stages of enterprise adoption of the Internet to today's complex needs for network security technologies.

V-ONE is developing virtual private network security products for technology-sophisticated users with complex information sharing needs who are today's leaders because they are defining the outside edge of the secure communication envelope. These customers demand capable security solutions that take full advantage of private network communication technology and the Internet. This leads to the definition of requirements that ensure communication, information sharing, and protection for that information regardless of where and on what platform or device their information may be traveling.

What we are hearing from our customers is that they need to and will be on the leading edge. While their networks were not, in many instances, originally designed and developed to share information, they are now being confronted with the absolute need to communicate and securely share information. The security technology must provide them with vehicles to securely communicate and a security mechanism that allows different owners of data to open their networks and share the information, with assurance that the integrity of their systems can be trusted. If they do open their network and allow information to be shared, they, as the owner of that information, want to control all access to their information, provide access permission only to the level to which they want a particular user to have access, and know who that user is and that they can authenticate beyond any reasonable doubt that the person in cyberspace representing himself to be a certain person is in fact that person.

V-ONE vs. Other Companies

The patented software design of the V-ONE products is quite different from many of the products that are provided by our competitors. Many of the competing security products are built on an industry standard called IPSec, an Internet security protocol that is implemented at the network layer. It provides an encrypted tunnel, or a safe pathway, for data to travel across the Internet as it moves from one user or network to another. When passing information through a firewall that protects the perimeter of an organization's network into the Internet, data is placed in this protective tunnel in an encrypted form so it can flow safely to its destination.

While network-based IPSec VPNs provide data protection, they offer little flexibility for implementation of secure information sharing across organizations and with business partners in a distributed, collaborative configuration. The V-ONE product architecture offers data protection technology at both the network and the application layers. By offering the alternatives of IPSec and operating at the application layer, V-ONE solves the problems associated with network-layer IPSec-only deployments.

IPSec deployments require an understanding of the architecture of the network that will receive the data, making it very difficult for IPSec products to be implemented and deployed in an environment where you do not control the receiving network. With an application-layer product, which is the core design of the V-ONE security products, you do not need to know anything about the receiving network. All you have to do is protect access at your own network with your own VPN server, and the communication can reach the intended partners very effectively. It avoids all of the difficulties and problems the industry has struggled with for the last ten years in trying to get broad deployment and communication where only one side of the network is under the control of the IT administrator.

In many of the products in the marketplace, using an IPSec-based technology will take more time and effort for the initial installation. Network engineers have to be involved and communicate with each other to do the initial setup. And, if you are dealing with a business partner, you will not necessarily know or be able to authenticate the individual user who is the sender. You are dealing with a device, a tunnel, and an IP network address only.

If you need security at a level that allows you to collaborate, where you want people to be able to provide information to you over the Internet, the data encryption algorithms, or the security component, to a certain extent is the easiest part. They are well known, and once they are tested and certified, the encryption algorithms for most products will provide an adequate level of security. It is the administration, the management tools, and the ability to authenticate to an individual level that discriminate the products. For an organization to allow users from outside organizations to access and share information stored within each organization's private databases, the authentication, management and access control components of the security software, in addition to the encryption, must protect the communication link and the integrity of each sharing organization's information and must be assured by the receiving organization. V-ONE Corporation's distinctive SmartGate® technology allows an organization's IT manager to define, manage, and control access by individuals or by groups and can limit an individual's access to only certain specified information to be shared with authorized individuals, ensuring security of the data on the network. The owner of the information maintains this level of control at the network access point.

V-ONE products support user requirements for data encryption, user identification and authentication, precise access control, and audit logging, and they do so for a broad range of system and user environments. Encryption algorithms including Triple DES and AES are combined with the products' powerful V-ONE integrated authentication token that has been validated to strict U.S. government standards. Support for a wide range of third-party user authentication systems can also be seamlessly incorporated into the security architecture if preferred by the user. V-ONE software is available for multiple operating systems and multiple end-user devices. If you happen to be in an environment where you can use your own desktop, that is fine; and, if you are traveling with a laptop and connecting from someone else's environment, that is also possible with the V-ONE technology. If you are carrying a PDA and transmitting wirelessly, that also can be done in a secure mode that resolves the fundamental insecurities of the 802.11 wireless LAN protocol. If you need to work with a pager for two-way messaging and email, the V-ONE technology can also provide that level of security. Broad user device reach and precise control at the server give the owner of the information the management and access control protections they require.

The result is that V-ONE technology enables people using our security products to be extremely fluid, highly flexible, and untethered – they can communicate securely from almost any location over wired, wireless, and satellite networks.

Setting Up Security Plans

One of the things we do routinely – this goes back to our perspective of the industry being fundamentally about people – is ask people what they need their system to do. Our preference is for them not to think about security *per se,* but what risks they perceive and what protections they want to achieve. The V-ONE products satisfy both internal and external security requirements. We have security products that are inside a closed network because a particular customer, even though they had no access to the Internet, was very concerned about internal security and wanted the system administration and management tools for their own IT manager to provide access permission to their own employees inside a closed net. We need to know what the customer wants to accomplish and what their priorities are.

At V-ONE, because we have developed code continuously for more than ten years, we can provide security technology for all of those aspects. Our customers look to V-ONE, in most cases, for its real strengths in extranet technology – sharing information, communicating with business partners, collaborating through multiple devices, and working within regulated industries where compliance with security and privacy issues makes the difference as to whether communication across the Internet is possible and where identification to the individual user level is important.

Building Trust in New Customers

When we work with a new customer, one of the first things we do is demonstrate the technology. We take them through the validations and certifications we have received. Customer references are very important to add credibility. We have a list of installed customers and a portfolio of case studies. For companies that want to speak to customer references, we introduce them to customers that have worked with the company. We are installed in many very large networks and have customers who will speak on both the integrity of the algorithms, which are certified, and the ability of our products to work effectively within their own environments. We try to match the prospective customer with a customer who has many of the same needs, so that they can communicate about how things are handled and how problems are solved.

Most enterprise networks in today's environment are designed and implemented differently, frequently using non-standard elements, and the security implementation must fit the network design. It involves a lot of customization, and once the customer is comfortable, and the product itself is certified, and the algorithms are appropriate for the level required, and the application software will work, the most important thing is for them to be comfortable that they have a responsive company that will stand behind the product. They have to know we have the talent and the skill to help them through a problem that might have nothing to do with the design, architecture, or functionality of the security product, but in fact something to do with their own applications, firewall settings, or network design. As customers continue to evolve and create sophisticated network implementations, they also create additional customization and complexity issues that have to be solved. We keep a team of system engineers that not only know our own products very well, but also have the security and network expertise to go onsite and make sure things go smoothly.

Continuous product development, onsite product demonstrations to answer the technical questions, and customer references to answer the questions about customer support before, during, and after the security product installations differentiate V-ONE's security products.

V-ONE gets high marks from our customers on all fronts. It is our culture.

How Do You Know You've Developed Successful Security Products?

The criteria for a successful security product are twofold – whether it works as it is supposed to work and whether there is a customer base willing to buy it. The products have to function and perform the required security protections the customer needs in their own environments, and they also have to have the flexibility and ease of deployment that is necessary to get them rolled out.

Much of our research and development process is a customer-driven cycle. We work with government programs and offices that reach beyond where the industry is today. To a certain extent, this is driven by the tragic events of September 11th. The government acknowledged a new requirement and stepped into the role of a leader in designing security for information resources they wanted to be able to access and share. The difficulty for these customers is how to put something together today with existing systems while they also build for tomorrow. Our products are designed to help them do just that. As a company, we are now evolving a sophisticated legacy product and can, through the requirements of our customers, create new products, new functionality, and new capabilities that keep the V-ONE technologies on a leading edge.

From a commercial business perspective, we have been able to move what is being developed for government entities very effectively into the commercial marketplace. All of the systems rely on telecommunications in one form or another, and developing security products to meet unique customer network designs is not inherently different. The information-sharing requirements will be the same, whether we work with a government agency or a commercial business.

Generating Revenue

V-ONE is a product company, and the majority of our revenue comes from our security product sales and service revenue for product maintenance and enhancements.

Much of what we do that is also an additional generator of revenue is to provide support to customers or to integrators for security and network design, which is to a certain extent an evolution of the scope of our product business.

We also provide training and education on the different types of products available and create case studies and snapshots so that potential users or purchasers of security products will know there are differences. It becomes almost an education component, and that education pushes itself into something that is also a revenue generator.

Biggest Challenges in the Marketplace

The knowledge level of CIOs and CEOs responsible for making the decisions about access to the Internet and selecting security protection is still not where we know it must be to make informed choices. Many people are just beginning to look at using the Internet for cost savings. Some people are still in the mode of completely restricting their internal assets from any access to the Internet at all.

The biggest challenge we face is the realization that we are in an emerging and evolving industry and that customers are still in their learning stages. The level of experience and knowledge of potential customers still often gate realization of an adequate ROI for security product providers. For V-ONE to be competitive in our markets, we have to be appropriately compensated for the training, engineering consulting, network design, and the entire services component of the product offering. In many cases, we have to put people onsite, and it is difficult for a new customer to understand there is a cost aspect there in addition to the product cost.

The other significant challenge in the marketplace is staying ahead of security demands. In the customer environments we serve, we provide products that have to work with and support other organization's or vendor's products. We provide security software for multiple operating systems. Our server works with Microsoft, Solaris, Linux, and MAC. Each time these manufacturers create enhanced applications or changes to their operating systems, we have to upgrade to stay in step, working with them in early release or, in many cases, on a beta release to make sure they have not opened something in their application that may have been closed in an earlier version. In this early-stage industry, minimizing vulnerabilities is an element of security product development and quality assurance (QA).

One of the things we understand in developing security products is that an operating system is a very sophisticated program that has to be designed to allow data to flow to and from applications. All of those applications are open, and they talk to each other. To embed a security piece around that is an extremely difficult, if not impossible, task.

Many applications and operating systems have that same architecture. They are designed to work inside a network and have to allow the free exchange of data flow. Security products are designed to capture data as it flows outside of an operating system or application and protect the data during its travel. If there are changes in applications, the next version may not interact with the security product in the same way or the same place. A security product must check all changes to be sure a new application design did not open something you did not know about. You hire hackers; you hire testers; you work with beta sites; and you cooperate and collaborate on early releases to make sure that when a new network application product reaches the market, if you have a security product to protect it, it does in fact protect every aspect of the new release. That is one of the most difficult challenges any provider of security products will face because what was a current release yesterday is obsolete today. We have to constantly be on that leading edge to make sure we have not missed something.

A serious situation would clearly be something in the software release that did not get caught during the QA cycle and that would cause a breach or a problem for customers. V-ONE has gone through ten years of development in which we have instituted best practices that keep our product as solid as it can possibly be in this fast-paced emerging industry.

When security is your business, you have multiple layers of QA that your products go through, including a beta-testing process with some of your most comprehensive customer users. These users allow you to put a beta product onto a restricted part of their network where they will run it through some serious paces to make sure there are no holes, helping you in the testing cycle as you are ready to go to final release.

One of the things we have also done with our products is to purposely hire white-hat hackers and let them try to break the code.

The other requirement, because we do sell into the federal government, is to provide our encryption algorithms to the National Institute of Standards and Technology for validation.

Changing Times

As companies or government agencies go through the decision process on how they will spend their money, the arguments we often hear about security products are the same kinds of arguments that would be given to an insurance company. Security is insurance, and the cost benefit tradeoffs are whether or not the information has to be protected, how much of it has to be protected, and at what level.

In good times, many of the decisions would be much broader; decision makers would have a more comprehensive sweep of an organization to put the security in place for everyone. In a down economy, where everything is tighter, they will look more at closing up the networks and providing security access to only a certain level or a certain business segment or component. They are becoming aware that the security is required, and if they are going to open any aspect of their network to the Internet, protection is necessary. If they are going to open a critical design element of their business to communicate across the world via the Internet, they will have some level of security, but it may not be available for every individual on the team.

Security will be taken through the budgetary process of any organization and require business justification for expenditures. The critical processes of the business to keep its doors open and operating will have first priority, and even though the breach of a security system could cost them the business and their livelihood, it takes second place at this point.

Planning Strategies for the Future

V-ONE is a product company in the security industry. Our products are protected by eight patents and offer a very comprehensive security solution for mid-sized to large enterprise organizations. We benefit from being in development for more than ten years with a product that has evolved its original design functionality to serve complex customer requirements, and the marketplace is now embracing the products because of their comprehensive scope.

The industry standards body designed and defined the IPSec protocol almost ten years ago. V-ONE's product vision was different. It focused then, and it focuses now, on how the individual uses the Internet and how the product can service that individual's security needs. The industry has learned a lot since then, and so has V-ONE. The way V-ONE's products were designed was the application layer as its first design criterion. The industry is, through its own evolutionary cycle, recognizing the importance of application-level security, and the industry is now calling for V-ONE's vision in the products it is adopting for today's information sharing and collaboration needs across the Internet. V-ONE has included IPSec solutions in its product suite to provide the customer with a single product, one license cost, and choice as to which protocol to use in this complete security solution.

Our strategy is to focus on our customers, continue developing products that take advantage of our patents that protect the design and methods of application-level security technology, use IPSec where this protocol is most efficient and enhance system administration and deployment capabilities, and continue to stay right on that leading edge.

We stay very aware that as the industry becomes more knowledgeable, consolidation will eliminate the technology outsiders, and requirements definitions will encourage security products to become more mainstream. Security implementations will be seen and understood as part of government and industry policy and practice, and the type of security that will incorporate both capabilities – application layer and IPSec – will be in demand.

Because we have already reached mainstream capabilities ahead of many of the competing products, we need to very carefully use this next window of opportunity, where we have a technology lead, to make sure we stay ahead. We have to pay close attention to the signals. As the industry moves along, the product life cycle will change, and security technology innovation will become a different challenge. The additional product evolution, which today encompasses many technologies – including intrusion detection, network forensic tools, firewalls, VPNs with application layer, SSL and IPSec support – may be combined into one solution through product evolution and consolidation. We see this as a ten- to 15-year cycle.

Major Changes in the Future

The industry, as it moves forward, will see people quickly gaining knowledge of the real power of the Internet, and the ability to take advantage of and push information technology in a way that will give their business or an aspect of their interaction a competitive edge.

With that, we will deliver networks and communication security tools as quickly and as powerfully as we can to our customers' advantage. Security will not be left behind. Education and awareness are establishing this basic need today.

This industry will move more rapidly in the future than it has moved in the past because there are so many people now who have reached a level of basic understanding, and with that fundamental base under them, the ability for them to take the next step has been enhanced tremendously. That means the speed with which they will take that next step is faster than we can imagine. Making sure that next step is taken in a way that embeds or incorporates security as fundamental, as something they no longer have to think about, has not happened yet. It needs to happen and will, probably over the next five years.

The industry is still quite young, and people are learning rapidly. Applications developed first, and security followed. The door was open to the Internet before people understood what it meant, and there are more people involved in the business of business than in the business of security. Vulnerabilities and the risk/return tradeoff will drive security as a business enabler.

Golden Rule of the Security Business

For security, the golden rule is that people's safety, protection, and privacy are fundamental aspects of human nature. If the security products address those fundamental needs, and customers understand that the best practices will involve creating an environment where trusted communication can happen, then even in cyberspace, people will be able to look eye-to-eye and conduct their business in a manner that has been done since people began doing business.

The industry is evolving, growing, and maturing. Many competing products will come into the marketplace. As people make decisions about what is right for their own environment, it will become more important for them to take advantage of and learn from best practices, avoid mistakes that have been made in the past, and then build their security policies and practices around what makes sense for them. Every set of needs is different. They can't skip the best practices of good business, and they must do their own internal brainstorming and set their security parameters. The organization has to begin to think of security as their responsibility, educating themselves to the risks and the available products and solutions to make sure they don't expose their network to harm themselves and that they don't allow their network to be used to harm others.

Best Piece of Security Industry Advice

The best piece of advice I have ever received is to make sure that if you make a promise, you deliver on it.

You don't always have to be first; you don't always have to be the leader; but you must understand what you do and do it well. You can carve out a very important place in a market if your product is solid. As important as the security products we deliver are the trust and confidence our customers have in us. It all comes back to trust.

Margaret E. Grayson is the president and CEO of V-ONE Corporation, a network security pioneer. V-ONE offers a suite of enterprise-class software products and hardware appliances that provide a complete virtual private network (VPN) solution for V-ONE's customers. Fortune 1000 corporations and sensitive government agencies worldwide use V-ONE's innovative technology for both wired and wireless integrated authentication, encryption, and access control.

Before joining V-ONE, Ms. Grayson served as vice president and then CFO for SPACEHAB, Inc., and chief financial officer for CD

Radio, Inc., in Washington, D.C., an early entrant in the satellite radio mobile communications market. Previously, Ms. Grayson served as a senior executive and consultant to high-technology start-up companies. She was principal financial advisor for raising private and public financing, investor relations, structuring and negotiating joint ventures, and completing five successful acquisitions, both domestic and international.

Ms. Grayson is a member of the National Infrastructure Advisory Council (NIAC), serving at the request of President George W. Bush. She is on the board of directors for the Montgomery College Foundation and the Dean's Advisory Council for the School of Management at the State University of New York at Buffalo. Ms. Grayson holds an MBA from the University of South Florida and a BS in accounting from the State University of New York at Buffalo.

Ms. Grayson has published a number of articles on security and the protection of cyberspace, and she is a frequent speaker on such topics as corporate financial management, enterprise network security, and government law enforcement information sharing for homeland security.

Emerging Security Risks and Solutions in the Digital Era

Bruce Davis
Digimarc Corporation
Chairman & CEO

Needed: Balance Between Security and Liberty

Concerns over identity theft and counterfeiting have increased to unprecedented levels because of a combination of many factors, including increasing threats of terrorism, widespread commercial fraud, and illegal claims of citizenship and entitlement. These motivations for identity theft and fraud are translated into false credentials with greater ease as computers and software permit even those with modest technical skills to simulate the security features of identity and value documents. These trends create needs for new, more effective deterrents to protect a wide variety of documents, from banknotes and checks to driver licenses and travel documents. Increasing insecurity is not limited to value and identity documents. The same advances in digital technologies that are at the root of these problems are contributing to music and movie piracy and trafficking in counterfeit and grey market goods in the United States and abroad.

Global traffic in people, goods, and media has become very fluid and anonymous, contributing to these problems. Security systems need to evolve ways that enhance personal and economic security with minimum impact on liberty, privacy, and effective commerce.

21st-Century Security Vulnerabilities

Individuals and companies are increasingly vulnerable to security breaches. The attacks are becoming more sophisticated. As we build more complex networks, paradoxically more opportunities for insecurity are created. As people begin to shop online and as more people travel, they become more at risk, and the security becomes greater in both scale and complexity.

Our company addresses increasing security risks due to changing technologies and business processes, and increasing scale of operations, by conducting regular audits of our networks and physical security. Being a security-oriented company, we are more conscious of these issues than most. Despite our relatively small scale at less than $100 million in revenues and about 400 employees, we have begun considering creating a new executive position of chief security officer. The consideration is motivated by increasing complexity and relevance to business stability and growth of the issues involved.

In the old days, a network administrator or a CIO would be in charge of the network security, and there might be someone in charge of physical security in the facilities department. Now such matters are increasingly interrelated.

We routinely do criminal and financial background checks on everyone we hire, a practice becoming increasingly common among medium to large corporations. All exterior doors are secured with contact-less smart-card locks. Interior spaces are likewise secured. We have video surveillance in various areas of our facilities. We have not yet concluded we need to restrict portability of computers or Internet access, although we stand ready to do so if a threat to business security is identified.

We have a great deal of network security, including virus protection, firewalls, intrusion detection, virtual private networking, hierarchical privileges, backups, offsite storage, password protection, and segmentation. We educate our employees about how to maintain the security of their workstations and network connections.

Management of enterprise security is increasingly complex, ranging from proper identification of new employees and making sure the people we hire do not have an inherent security risks to maintaining the integrity of the facilities, workstations, and networks, controlling the flow of information, and making sure our employees have access to mission-critical information only to the extent they need to perform their duties.

Security Technologies Are Inherently Imperfect

There is a natural tendency in difficult economic times like the past few years for management to not want to invest in security beyond areas in which they have already suffered vulnerabilities, as they balance the costs of such measures against risks amid severe pressures on short-term profits. These are risky judgments. All too often in security matters, people experience a loss, and then lock the barn door after the cow has escaped. The significant loss precedes spending on measures that would have prevented it. The loss vividly demonstrates the risk that was being tolerated. Obviously, everyone would have been better off if cost-effective security measures had been in place previously.

Perhaps the highest-profile losses from digitization of the economy are from music and movie piracy. The music industry has suffered devastating losses. The movie industry, seeing the writing on the wall, has an opportunity to learn through the losses of this analogous industry and put measures in place to avoid a similar fate. They have been engaged for several years now in technological studies, legal analyses, and lively debates with consumer electronics and personal computer industry leaders about potential solutions and whether governmental intervention is required to implement an effective and timely solution. Digital watermarking is one of the key technologies being discussed.

I would say to a potential purchaser of security solutions that if there have been avoidable losses and similar situations, they should think about how their business will be developing. If they anticipate those circumstances that put them at risk, like those that have already caused losses, they will understand that they should invest now to reduce the risk of those losses, rather than wait until it is more obvious and more costly.

Probably the biggest misconception around general management, CEOs, and other executives is the notion that if security does not provide 100 percent protection and is not hacker-proof, it is not worth buying. Often you will find people in the market who say they heard the protection was breached. They are missing the point. The point is not whether the protection was breached. The point is whether the user saved more by having it than by not having it. It is an economic judgment, not a binary judgment of whether it works or does not work.

Despite best efforts of all involved, insecurities will persist. It is important to recognize that all security solutions are imperfect in that they involve many compromises and can be subject to unforeseen attacks. The value of security solutions is a function of a complex equation, weighing financial and operational costs against the effects on costs, expertise, visibility, and time necessary to overcome the measures. Generally, security-minded customers will implement solutions where the marginal cost of the solution is less than the marginal benefits in enhanced security that it provides. Frequently in general media, security breaches that are characterized as "failures" may be quite cost effective. Generally, security is a game of cat-and-mouse with the attackers, with continuous improvement an important aspect of the overall solution.

Motivated attackers study systemic vulnerabilities. Wherever efforts are undertaken to strengthen a system because of a breach, the attackers move to another component or subsystem where they sense more vulnerability and attack there. We see this most abundantly in Microsoft Windows, as people attempt to hack it continuously. It is a large, complex system, and thousands of hackers around the world continuously probe it for vulnerabilities.

To stay a step ahead, security solutions providers often employ or mimic the mindsets of the attackers. A good security design encompasses as much contemplation of how hackers would attack the system as possible.

The decision as to which security solution to adopt is specific to the circumstances of the particular customer and the risk they are trying to mitigate. A wise and well informed potential purchaser will quantify or otherwise characterize the economic risk or social risk, hypothesize the most likely and most serious attack scenarios, and estimate the costs of mitigating the attacks and the savings associated with mitigating the risk. It becomes an actuarial exercise like those done by insurance companies because of inherent uncertainties in many factors.

Sophisticated buyers of security solutions seek to increase the costs, time, expertise, and visibility of the hackers' efforts so that enforcement resources can be applied efficiently and with maximum impact. The attackers essentially are funneled into a net where they can be captured or constrained.

For these reasons, continuously improving and developing new products is critical to the security solutions we sell. If we were to stop the continuous innovations, hackers would catch up and overwhelm our customers' interests.

Digital Watermarking and Secure Identification Solutions

Digimarc is an unusually innovative company. We are about seven years old and already have over 110 issued U.S. patents and 400 more pending. Innovation is very important to our business. The company has very imaginative and intelligent people who are constantly coming up with great ideas for improving the lives of our customers.

We are a leader in two related areas of business: secure personal identification and digital watermarking.

In the secure personal identification area, Digimarc is the leading supplier of driver license issuance systems in the United States. The driver license serves as the primary means of verification of personal identity in America. Driver licenses are used in many contexts other than proving competence to operate a motor vehicle. Licenses are commonly checked at airports, many large commercial buildings, and at commercial establishments selling liquor and tobacco products.

Digimarc provide the entire issuance system used to produce driver licenses to state departments of motor vehicles. We have an intimate understanding of the marketplace built on longstanding customer relationships, many of which span decades. Product development in this area of our business is a cooperative effort with our customers and largely driven by collecting the requirements from the experience of supplying these customers, identifying areas for improvement in system design, and taking those improvements back out to the customers with proposals for implementation. The sales process is typically concluded after prospective customers include the improvement in one of their periodic requests for proposals or contract extensions.

In areas of our business other than secure personal identification systems, digital watermarking provides a means of persistently identifying media objects in a manner that is generally imperceptible by humans in ordinary use, but is easily recognized by digital devices equipped with our software. This embedding capability is applicable to all media objects, whether in analog or digital formats, including video, audio, printed materials, and digital images.

Today, for the most part, the identity is media-processed by computers and other digital devices not known to the device. A picture of an apple looks just as much like a random bunch of bits as does a news photo of an avalanche in the Alps. With digital watermarking, the computer can recognize the media objects and execute predetermined applications. For instance, it disallows copying of copyrighted subject matter unless the user has permission to do so. The application can also enhance the communication experience, for instance, by linking the user to additional information about the subject matter of the media object or to a licensing site to obtain appropriate license rights online.

One prominent use of digital watermarking for content security is by central banks to deter the use of personal computers in the counterfeiting of money. When someone tries to counterfeit banknotes protected with digital watermarking and related technologies on a computer, the computer refuses to process it because it recognizes the media being processed as a banknote.

Increasing Influence of Technology on Security

Security is a broad field. There are many security solutions, ranging from security guards to private investigators to intrusion detection on networks. The industry has gained notoriety and increased demand

and investment during the last few years. A decade ago companies like Pinkerton and ADT would be top of mind. Now, we add technology companies like Checkpoint, Macrovision, and Symantec to the mix. Security has gone high-tech.

Digimarc is upping the technology quotient in the markets in which it operates. For instance, in driver license fraud, a common problem is fraud in enrollment. In some jurisdictions, people can go to the DMV with a couple of pieces of mail or a credit card or some other indicators of identity, prove they can drive with minimal competence, get a driver license, then use that license to travel, establish credit, etc. We have begun to develop sophisticated software tools that are aimed at reducing the fundamental risks at enrollment that trickle throughout successive iterations of establishment of identity that are built on the driver license credential. One such solution was deployed in Colorado in late 2002, involving the use of facial biometrics to match new applicant photographs against all existing licensees in that state to determine whether the applicant had previously obtained a license using a different identity. Colorado officials report finding frauds nearly every day using this tool.

Viewing secure personal identification in a more global sense, what we do in driver licenses as a sub-system of a network of secure personal identification is of increasing importance to personal security and facilitation of travel and commerce. There is a great deal of progress ahead of us.

Generating Profits, Thriving on Change

Digimarc has an aggressive, growth-oriented management team. Our company has experienced more than 200 percent average annual growth since its inception. We are fostering organic growth while actively pursuing acquisitions that support our mission and financial objectives. Acquisition has been a big factor in our success: We acquired the assets that spawned Digimarc ID Systems from Polaroid Corporation in 2001. ID Systems accounted for nearly 90 percent of revenues last year.

When we started our company in the mid-1990s, security was a niche. Since then, it has become front-page news and a much larger industry, with a wide range of products and services and much more technology. Despite the scale, there is quite a lot of fragmentation. These are ideal conditions for entrepreneurs: many relatively small suppliers, imperfect information flows, unsettled economics, evolving perceptions of needs, and rapidly ramping demand.

The Future of the Security Industry

Market demand will drive greater interconnections among solutions, leading to consolidation of suppliers. Customers will become more sophisticated and comfortable with the investment spending, imperfect solutions, and unquantifiable risks inherent in this field. They will think more broadly about risk management and total costs of implementation, including financial and process costs, versus risks and costs of business and societal disruption. Risk management will become a more obvious aspect of everyday life as people seek ideal balances between security, lifestyle, and work style – and the costs of achieving and maintaining those balances. There will be no "one size fits all," as many varied factors will be accounted for in the equation.

Our core digital watermarking technology provides an inherent imperceptible digital identity for all media objects – movies, music, digital images, and printed materials alike. We believe this technology can have extraordinarily broad relevance in security solutions. A number of such solutions have already been deployed. Many more are being developed by Digimarc and its licensees.

The promise of digital watermarking will continue to unfold over the next ten to 15 years. Identification and effective management of media objects will be central to 21st-century security solutions. In some cases, like movies and music, the intrinsic value of the object and new threats of unauthorized digital processing demand new means of control. In other cases, like driver licenses, protecting and enhancing extrinsic value as a means of authenticating the holder is the motivator. Whether the principal value of the media object is intrinsic or extrinsic, persistent digital identification of the object will be a key attribute of its usefulness in the digital era.

Bruce Davis has served as Digimarc's chief executive officer and director since December 1997 and was elected chairman in May 2002. He also held the title of president from December 1997 through May 2001.

Before joining Digimarc, Mr. Davis was president of Prevue Networks, Inc., the leading supplier of electronic program guides and program promotion services to the cable and satellite television markets. Prior to that, he founded and served as president of TV Guide On Screen, a joint venture of News Corporation and TCI that supplied electronic program guides and navigational software for the cable television market, and was merged with Prevue Networks. A pioneer in the development of the video game industry, Mr. Davis served as chairman and CEO of Activision, and served five years as a member of the Software Publishers Association board of directors before beginning his pioneering work in TV program guide development.

The holder of more than 20 patents, Mr. Davis is well acquainted with the need to protect intellectual property. Mr. Davis started his career as an intellectual property attorney. He holds a JD from Columbia University, and both a BS in accounting and psychology and an MA in criminal justice from the State University of New York at Albany.

Improve Security with Identity Management

Kurt Long
OpenNetwork Technologies
Founder, President & CEO

Basics of the Security Industry

With the growth of e-business, companies are faced with the challenge of managing access to information and applications across internal and external networks and for thousands — even millions — of users. To be successful, organizations must improve their ability to manage user identities, provision their use of company resources, and control access to those resources. This is known as identity management.

OpenNetwork Technologies is a software company that makes it possible for organizations to meet the challenges of identity management. OpenNetwork's products provide authentication, authorization, privacy, access control, user management, and auditing capabilities to help large companies better secure and manage their employee, customer, and partner populations.

At the most basic level, solutions like OpenNetwork's ensure that users are who they say they are (authentication), and that they have permission to access a requested resource, such as a Web site or other protected resource (authorization). These capabilities help guarantee privacy and ensure that no one has access to personal information other than authorized parties.

At the next level is the identity management functionality that enables user and security policy administration, self-service for things like password reset, self-registration for online resources, delegated administration, and auditing. Auditing provides a business with the ability to review historical data and learn who has accessed and used a resource. It enables companies to understand the value of their internal and external applications and to meet the strict requirements set by many regulatory agencies. These capabilities help companies realize administrative cost savings, provide better security, and implement improved business services across their global communities.

Attracting Customers

Outside of traditional marketing efforts and strong partnerships with leading consultants and resellers, it is imperative for software companies to make a solid business case to attract customers. In an era of strained budgets, technology infrastructure investments need to be justified with tangible benefits and ROI.

The following outline is a simple approach to creating a compelling argument for a technology purchase:

Step 1: Determine the business pain and define the value of the technology that will eliminate or mitigate that pain. It is critical today to enter a prospect meeting knowing that the security product you are selling is not an extraneous piece of hardware or software, but can actually help a company achieve strategic business goals like improving productivity, reducing IT burden, and lowering costs.

Step 2: Establish credibility with the customer or prospect and illustrate various approaches to long- and short-term development options. In other words, clearly spell out the ways that the technology in question can help today and grow to support the customer needs in the future.

Step 3: Provide a better product and better service than anyone else in the industry. Focus on R&D and quality control to ensure that the solutions sold are the best they can be. In the technology industry there is a tendency to ship a product quickly and fix problems later. I believe the most important part of my job is to demand that the product is as perfect as it can be when it leaves my door, and any revisions only make it better. It is also critical that the company culture is one of customer centricity. A company focused on customers attracts and retains them.

Corporate Vulnerability

With the litany of viruses that travel through the Web every day and the always-possible threat of hackers, corporations are vulnerable if they do not take the necessary precautions. To mitigate risks, security technology must be implemented at every level of the network.

I think of security, particularly identity management, as an enabling vehicle for communication and commerce. Our product provides a security layer between a person and valuable information, and it allows access into trusted environments. When customers, partners, and employees know that they are in a secure environment, they are more likely to shop and conduct critical business online.

Additionally, identity management prevents losses associated with identity fraud – external theft, internal theft, or the compromise of sensitive information that can result in lawsuits. If technology managers are savvy and understand the best products and the most effective ways to use them, they will be able to protect their valuable corporate assets and minimize risk.

Security Issues for Businesses

Any company dealing with critical and sensitive information needs to carefully consider the security infrastructure they put in place. Our client base consists of global organizations with financial, human resource, regulatory, and many other types of data. They need to provide access to information used by growing populations of employees, partners, and customers across the world. With this type of critical material and a large audience, there is an immediate security need and a possible threat.

Our software ensures that corporate resources are protected and accessed only by those authorized to do so. Here is an example of a process that may occur – but can be prevented.

An organization grants network and application permissions to Employee A through a process called provisioning. That person is responsible for managing global, financial, and regulatory transactions for a multi-national customer and employee base. One day, Employee A has an argument with a senior manager and resigns from the company. Without the right solution, Employee A, who is now a disgruntled former employee and potential threat, can still access all of the company's resources.

With identity management software, Employee A's network and application permissions are immediately removed, or deprovisioned, from all systems, preventing unauthorized access and reducing the likelihood of a security breach.

Enabling Client Objectives and Building a Security Plan

When we engage with a client, we are often called on to put together a security plan. There are several areas we take into consideration before determining our best implementation approach.

First, we meet with the business decision makers to gather historical data and understand the overall corporate strategy and business goals they are trying to meet with the product. Next, we speak with the development teams to gain a greater understanding of the network environment, its strengths and weaknesses, and the areas we can improve on with our solution. Joining the business strategy with the tactical goals, we put an achievable plan in place that will satisfy the needs of our internal and external audiences.

I can't stress enough the importance of aligning business and technology goals with the deployment of a security solution that will benefit all facets of a corporation.

Making a Security Product Successful

Deployability is key to making a security product successful. OpenNetwork believes that our product, Universal IdP, is successful only when a customer is using and benefiting from the solution in a production environment. As a company, our most important mandate is to get every customer into production quickly, without interrupting any of their business processes.

Of equal importance is the ability for a security solution to scale to suit growing business needs. For instance, even though an OpenNetwork customer may purchase a solution for very specific departmental functions, we always ensure them that our product can expand to support companywide deployments as demands require. Because we have dedicated development teams that are continually in contact with our customer base, we can provide the kind of customer satisfaction and success that is rare in our industry.

Challenges in the Marketplace

Some of the challenges that enterprise security companies face in the market involve the breadth and scope of delivering an enterprise security solution for a global business that operates on multiple continents, and the continual evolution of technology and standards adoption.

We work with companies like PricewaterhouseCoopers, Unisys, and Microsoft, to deliver our software and ensure that customers are maximizing the product across all their locations. Our partners have consultants in almost every country who assist in deploying our software and user training.

To meet the technology and standards challenges, we stay at the forefront of innovation. Our engineers are active authors and participants in standards committees and industry regulatory agencies, and we have partnered with some of the leading companies in the world (customers and partners) to ensure our solutions deploy easily and efficiently.

Continuing to Generate Profits and Revenues

The ability to constantly generate profits and revenues is a twofold process. First, you have to get your customers into production and make them successful. You need to ensure that when a customer buys your software, they derive value from it. If they derive value, they will then pay for support maintenance on an ongoing basis. Moreover, as their business needs warrant, they will purchase more software from your company. Basically, if you make the customer successful, the profits will follow. That is our core company philosophy and at the heart of everything we do at OpenNetwork.

The second part of the process is to constantly examine the marketplace and make improvements accordingly. By relentlessly evaluating our product against market needs and anticipating trends, we find ways to make solutions that are faster, better, and less expensive. For businesses to remain competitive, constant evaluation is critical.

Life Cycle of a Product

With any product, there is a starting point with the first customer success. Once the product has been initially sold, there will then be a revision of that product. At the revision stage, we listen to our various constituents to determine the requirements for our next release. We speak with our existing customers and talk to our partners. We then evaluate the competitive landscape on our own.

Based on these factors, we come up with a set of requirements that build on an existing product. Once we develop a prototype, we conduct unit testing, systems integration testing, and regression test cycles. We then release alpha or beta software to customers who provide feedback and assist in finding areas of improvement. After that process, we take a final improved product to the broad marketplace through our direct sales force, our partners, and our reseller channels.

That is basically the life cycle for a top-level product that has already experienced customer success.

The second kind of product development we conduct – and often the most critical – is to take a fresh look at the market to determine how we can better serve our customers and prospects with a future solution that would provide incremental value. For this type of development, we solicit input from customers, industry analysts, and partners. However, unlike improving a pre-existing product, we are starting with a completely fresh look at the market. Once the new product is ready for quality assurance tests, we put it through the same alpha and beta programs as our mature products. Those are the two primary ways that we introduce products to the market.

As we deploy solutions in our customer environments, we run our own regression testing for security, reliability, availability, and ease of use. In addition, our customers typically subject products to their own internal tests and enlist third party testing. That level of quality assurance makes the product increasingly better.

Putting Out the Best Product Possible

As I mentioned, OpenNetwork continually seeks to create a product that is more compelling that the last. Most importantly, the solution has to be deployable and easy to use and provide a seamless, transparent experience for our customers. It also has to provide a fast return on investment.

The smartest way to ensure product success is to continually listen to customers. There are many customers that we have a long history with and consider in some ways an extension of our product team. They and our partners evaluate our solution and continually provide feedback so that we release the best product possible.

Industry Conditions Affecting Revenues

With everything in business, when capital expenditure budgets are flush and global corporations are spending, it is generally a better environment for selling security software like ours. When capital expenditure budgets are tight, our revenues can slow down. However, that type of environment forces a company to focus sharply on the value provided to customers. In a tighter economy, organizations need to work more closely with customers to help them understand and articulate the potential value of a proposed solution. It is the difficult times and the creativity spurred from those times that set a company up for tremendous growth when times are better.

To manage a business like ours, it is also critical to manage with fiscal responsibility and to deliver products that can solve immediate needs for companies. This means that despite the economic environment, sales will remain reasonably constant, and the business will not be overly strained if there is a lengthened sales cycle.

Approach to Risk

As a software vendor, you always want to deliver a product that is exhaustively tested, comprehensively secure, and scaled to literally millions of users with an uptime of 99.99%. The risk is that someone is out there who will provide a product that is faster, better, and cheaper than your product. Competition is critical for excellence in a marketplace. It is good for consumers and good for businesses that have to push themselves to deliver quality products and services.

Another risk that companies, especially in the security technology sector, have to take is going out on a limb to introduce products that may be slightly ahead of their time. With the technology climate continuously changing, it is critical for vendors to be many steps ahead of the market and predict what it can support and what customers will demand long before they know themselves. To get truly innovative ideas, you have to be willing to take chances, have faith in your own business and technical knowledge, and be comfortable enough to operate in a bit of a vacuum.

Best Security Advice

When security is conducted properly, it is an enabling technology, so it is critical to always get it right. This was the advice from David Juitt, who was the former head of security for all of GTE business and was also employed at GTE Laboratories. One of our first customers was GTE Telephone Operations.

For clients, the best piece of advice I have given is to make security implementation a part of the strategic network planning. I have seen many businesses try to deploy applications electronically or reach a business objective in some electronic way without taking time to implement security methods. They get two-thirds of the way through the process of implementing an application, and things start breaking down.

By having jumped the gun and headed full steam ahead without considering security up front, they find themselves in a position where they will lose an incredible amount of time and money. If they had considered and planned to deploy the appropriate security technology at the outset, they would have the infrastructure in place and a lasting foundation for the next five to ten years, and could focus on achieving their core business goals, rather than having to face security as a roadblock.

My Personal Vision for the Future of the Industry

Identity management is one of the most compelling focus areas for enterprises. Based on the trends that we see in our business, the interest and need for the technology will grow significantly in the next five to ten years. Deployment of corporate identity and permissions infrastructure will accelerate, and next-generation e-business needs will evolve, causing vendors to offer more full-featured products that support a growing number of standards currently in development.

Key business drivers for the future of identity management will be Web application and Web services deployments, regulatory requirements, and evolving business strategies and goals.

One of the major developments that will affect the future of identity management and e-security is Web services standards, including the ratification of languages like the security assertion markup language (SAML) and the evolution of something called federation. Federation makes it possible for access management to span diverse corporate boundaries. In a federated environment, a user can log on to a home domain and then access resources transparently in external domains, including those managed by customers or suppliers.

There is no doubt that in the future, the Web will be a more convenient and secure environment for business users and consumers. Corporate applications across the world will interact with a behind-the-scenes protocol "handshake," and people will be able to traverse the Internet with one simple login that will allow them to transparently conduct transactions between corporate boundaries across the globe.

Kurt Long is the founder, president, and CEO of OpenNetwork Technologies. Founded in 1995, OpenNetwork improves the way businesses secure and manage Web and enterprise user identities and policies. Under Mr. Long's leadership, the company has become a dominant force in the enterprise security software market, particularly in the area of identity management.

Before founding OpenNetwork, Mr. Long held various senior sales and marketing positions with IBM. Early in his career, he was mission manager for the Space Shuttle Real Time Launch Databank at Kennedy Space Center, overseeing more than 15 missions, including the Hubble Space Telescope, Galileo, and Ulysses.

A member of the University of Florida College of Engineering Dean's Advisory Board, Mr. Long is also the chair of regional development for the Tampa Bay Technology Forum.

Mr. Long earned his MA degree in mathematics at the University of South Florida, where he has been an adjunct professor, and holds a BS degree in computer information science from the University of Florida.

Dedication:

My wife, Teresa Long
My sons, Trent and Hobie Long
My parents, Lloyd and Sylvia Long
The leadership team and employees at OpenNetwork
My first mentor from Kennedy Space Center, Rupert Stephens
My first security mentor, David Juitt
Friends and associates who have helped along the way – they know who they are

A Very Brief Overview of Active Security

Phil Libin
CoreStreet, Ltd.
President

The Challenge

Traditionally, security has focused on preventing bad things from happening by setting up checks, procedures, and roadblocks that only legitimate users could pass. Today's organizations must have security measures in place for all aspects of their business, but they cannot afford to rely exclusively on the traditional approach because it does not provide positive economic returns.

Threats to security are everywhere – both inside and outside national borders and corporate walls. It is no longer sufficient for businesses to focus security resources exclusively on the physical or electronic perimeters of their organizations. Digital firewalls and physical fences may be effective at keeping out certain types of external threats, but in today's increasingly interconnected environment, every important transaction must be protected. Well defined front lines and borders are becoming scarce and insignificant.

Now that security threats are pervasive, security can no longer be an afterthought for businesses. Companies cannot design their business processes and realize afterward that they had not adequately planned for security. True security must be designed into the fabric of every important process. The challenge is to determine how to make security more pervasive while keeping the costs and burdens of traditional security approaches from grinding business to a halt.

Can we make security ubiquitous, without making it overwhelming?

Today there is a new approach – active security. Active security enables every legitimate transaction to be better, faster, and cheaper. Active security does not hinder business – it protects while enabling positive business growth.

Two Paths

Traditionally, security was thought of primarily as negative or restrictive. It meant keeping adversaries out and keeping bad transactions from happening. Random screenings and searches at airports are examples of the old, restrictive approach. A percentage of all travelers are subject to some kind of search.

That approach is appropriate for certain situations, but the effect is to impede everyone's progress. Even though 99 percent of the people in line have perfectly legitimate reasons to travel, everyone is inconvenienced as the screening process is applied to all. This type of approach does not specifically target people who are the most likely to commit wrongdoing; rather, it pays a great deal of attention to every transaction – or randomly selected transactions – with the hope of being able to discourage or catch bad transactions. This restrictive approach prolongs the entire process for all.

Traditional digital firewalls take a similar approach. A firewall is simply a set of restrictive rules that say, "Do not allow transactions unless they are from certain individuals," or "Do not allow anything to be received from this port of that machine." The problem is that every time you add an application or new business process, you must determine how to make it work with the firewall. You need to design your application or process so that the firewall does not prevent legitimate transactions. The firewall makes it harder to write and configure legitimate applications, and as more and more applications get installed that are allowed to pass through the firewall, the overall protection afforded by the firewall dramatically decreases, since a bug in any of those applications may be exploited to enter the network.

The problem with a restrictive approach is that it screens every legitimate transaction, thereby burdening the overall economy.

Business margins tend to be relatively small and fine-tuned. If a business puts 10 percent of its budget into security right now, and that cost is forced to double because of a new threat, the entire margin may be wiped out. Increasing restrictive security is a very expensive option for companies; if they increase it too much, all business would grind to a halt.

By contrast, the new approach is to make security "active," or enabling. Instead of being restrictive, security technology needs to make every legitimate transaction faster and more efficient. With the active security approach, companies can prevent security problems and still conduct legitimate, productive business. If your technology is not solely focused on catching bad transactions, but instead is focused on making every transaction more efficient, then your security will both provide positive economic returns to the whole system and prevent security breaches.

Active Security in Action

An example illustrates the value of the active security approach.

A large publisher produces valuable and expensive reports for subscribers only – people who have paid for the service. The publisher's security challenge is to prevent unauthorized people from seeing these reports. How does the publisher make sure non-subscribers are forbidden from looking at the report, making a copy of it, or e-mailing it to a friend?

This is a traditional security problem. Under the old-fashioned, restrictive approach, the publisher would have to use some kind of technology that checks whether the person is a valid user, and if not, prevent the document from being displayed. This check inconveniences every user and makes it more difficult for a subscriber to look at the document. The publisher's customers are forced to endure the headache of either making sure they are registered on the machine, remembering their password, or having some kind of key. What's worse, this approach doesn't give legitimate users any positive incentive to cooperate with the security system. It's hard to keep out the illegal users if you can't motivate the legitimate users to help you. Making the system restrictive is not the best way to deal with security.

In contrast, the publisher tried an active and positive approach to protecting and validating both the user and the document. When a user tries to open a document, the system checks to ensure that the user is valid; if not, the document does not appear. This check is done efficiently, with minimum inconvenience. In addition, the document is checked to make sure it is legitimate and timely, that it has not been modified, and that it is the freshest copy available; otherwise, it warns the user. The subscriber might receive the following message, for example: "The intelligence report that you are looking at is authentic, but the authors have indicated that it is out of date. Do not rely on this being the latest copy."

Under this positive approach, subscribers have been given a distinct incentive for using the system. Now, it is a benefit to be a legitimate subscriber: The security system will tell the subscriber whether the report is timely and accurate, keeping them from making costly mistakes by relying on old data. Now, all the subscribers want to use the security system because it makes their business much easier. It also allows a company to install significantly stronger protection without generating complaints from users that the security is impeding their business processes. By motivating the legitimate users to embrace the new technology, it is much easier to keep adversaries out.

This positive approach to security is not science fiction. New technology makes it possible today to turn a restrictive situation into an enabling one. It is a new way of looking at security – less of a law-enforcement mindset for catching criminals and more of a business-enabling structure that makes routine transactions better and cheaper.

Traditionally Opposed Forces ...

The conventional wisdom is that several market forces pull security requirements in different directions, setting up a conflict or dilemma. These forces are traditionally described as *security, safety, privacy, convenience,* and *cost*. The idea is that to maximize one, you have to minimize another. Finding the perfect balance is tricky and time-consuming, so deployments are hindered. As is usually the case with conventional wisdom, there is an element of truth here, but the overall condition is overstated and misdiagnosed.

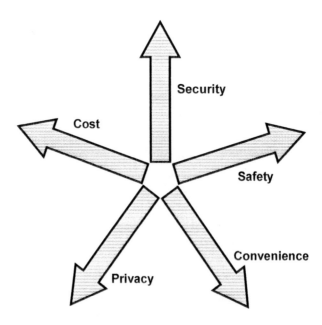

A conflict between "security," in the narrow sense of the word, and "safety" strikes some people as odd, but organizations have traditionally thought of security and safety as conflicting requirements. The primary goal of security was to prevent things from being stolen, while the primary goal of safety was to prevent bodily harm. If a company wants to install locks in its office building, for example, they can be designed to stay either opened or locked when there is a power failure or a fire. Locks that spring open in an emergency are typically called "safety locks" because they make it easy for employees to exit a building quickly. This is good for safety, but bad for security because someone could take advantage of an emergency situation to enter the building and steal data or property. Conversely, locks that default to closed are sometimes called "security locks." These are optimized for preventing theft at the cost of being poorly suited for emergency safety.

Convenience, privacy, and cost result in the same type of conflict. Traditional thinking is that if something is secure, it will be inconvenient. Searching every person at the airport is secure but inconvenient. It would be convenient for people to walk through unscreened, but the convenience would result in a lack of security.

Privacy is also often presented as a dilemma. Does the government have the right to know what a person is doing, or does an individual have a right to private actions? For example, should we require background checks on gun purchasers? Some people think it is necessary for society's safety and security, but gun rights advocates say the checks violate privacy and constitutional protections.

Cost is the last force seen as being in conflict with security. The traditional view is that the more security you have, the more expensive it is – both because of the direct cost of setting up security checks, and also because of the draining effect that restrictive security has on all transactions.

These conflicts are not actually inherent, but nonetheless have dramatically impeded the adoption of practical security technology. The current emphasis on new types of threats – combined with an active and positive approach to security – has shown that this traditional dilemma is no longer relevant.

… Now Pulling in the Same Direction

Theft is no longer the main reason to try to keep unauthorized people from entering a secure facility. A terrorist trying to defeat my corporate security system is probably not trying to steal something, but to kill and injure my employees. If my goal is to protect my employees from injury, it is not sufficient to keep them safe during a fire; I need to keep them safe from physical attacks. Security is no longer solely about theft. It is about physical threats. It is about safety.

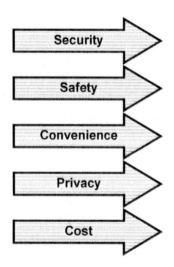

Similarly, a policy that inconveniences everyone makes people more likely to try to work around it, creating significant security holes. For example, many companies have a policy that requires all users to change their passwords every week. After all, if your password is stolen, the thief has only a week to use it before being shut out. But a rule requiring weekly changes causes people to forget their passwords. Their solution: Write the current week's password on a sticky note and attach it to their computer monitor – a security nightmare. The janitor, visitors, vendors, and everyone else who stops by the area can see the password. A policy meant to enhance security can so inconvenience people that it forces legitimate users to find a way to get around the system. The lesson is that we cannot have true security unless it is convenient for everyone. Those concerns – security and convenience – that seem to be in opposition are actually in alignment.

The same holds true for privacy. Identity theft is a concern as both a security issue and a privacy issue. If people cannot keep information private, identities will be stolen. If a thief steals the identity of a colonel in the military, the thief can presumably use the stolen credentials to get onto a military base and cause some damage. Those two concerns – privacy and security – have to coexist.

The concern of keeping costs low is similarly not in conflict with security. The assumption that additional security is expensive because you have to pay for each time your system detects a breach or catches a perpetrator is false for two reasons. First, consider the cost of *not* having security. The cost of one major attack may be an order of magnitude higher than the sum total of what you spent on security. Second, the active approach works so that security is profitable; it pays for itself. So those two concerns – security and cost – can coexist quite peacefully.

Security, safety, convenience, privacy, and cost – once seen as conflicting forces – are now closely aligned.

It's Not Easy Being Ubiquitous

Security has to be ubiquitous. It has to be everywhere, covering every transaction, and acting fast enough to prevent delays. This is a tall order. If a security process is activated every time a person enters a door to an office or opens an email or sends an instant message, it has to be fast. Security also has to be cheap. Adding a cost to every transaction is not feasible. It also has to be simple so that it does not inconvenience people.

Security was difficult to implement in the past because the prior generation of security systems was designed to function on the business periphery, as an afterthought. These systems were not designed to be part of every transaction.

Today's challenge is to find security that is up to the task of being everywhere: fast, cheap, easy to use, and scalable.

Bridging the Gap

Leading government and commercial organizations recognize that active security bridges the gap between physical and IT security. Many security publications are focused on this approach as well. Bridging the gap is important because if one side – either physical security or IT security – is weak, it can be exploited to do damage to the entire system.

Security Matters

In a recent, widely reported security breach in Australia, individuals dressed as repairmen entered the server room of a financial institution, spent an hour unbolting computers, and rolled a large server out of the room. The thieves weren't interested in the hardware; they wanted what was contained inside the server – highly confidential financial and identity information. The theft was successful not because the thieves had managed to bypass a firewall or VPN or steal someone's password, but because there was a weak link in the physical security at that facility. These men could walk in, say they had come to do some work, and leave with the computer – a prime example of a physical attack targeting data theft.

Of course, electronic security can also be breached for access to physical locations. For instance, the military in the United States issues Common Access Cards to every member of the military. These smart cards, once fully deployed, are supposed to be used for many purposes, including physical access to military bases. Loss or theft of a CAC must be reported so the card can be revoked. Once revoked, the card can no longer be used. But if a hacker can manage to infiltrate the computer system to "un-revoke" the card, that terrorist may be able to use the stolen card to enter a base and cause serious damage. This is an example of a virtual attack on a physical target.

The contrast between these two approaches – the actual incident in Australia, and the hypothetical base attack – shows that you cannot focus on *either* physical security *or* IT security. An attack on the weaker can expose both to harm.

Paying for Security

How much does security cost? The traditional, rather arrogant reply often used by security professionals is that the question is not whether you can afford to have security – it is whether you can afford not to have it. This response isn't good enough. It may be true that a business both cannot afford to implement appropriate security, *and* cannot afford not to implement appropriate security. The only thing left to do is shut down. That's not a very good answer.

Ubiquitous security – security everywhere, for every aspect of a business – would be crushingly expensive if it did not have positive economic returns. If security were required for every individual, discrete transaction, no one would be able to afford it. The only way security is affordable is by implementing it everywhere, which brings positive returns.

A security system that can enable processes like secure messaging and e-mail and the electronic signing of contracts enables positive business. It reduces overhead and makes processes faster and cheaper, eventually paying for the system itself.

The best approach is to invest in a security infrastructure that makes legitimate transactions faster and more efficient – enabling companies to make money and subsidizing the security infrastructure costs.

Take the example of the national ID programs being instituted in some countries in Europe and Asia. One country to announce such a program is Belgium. The Belgian government is going to make an investment to issue a smart ID card to every citizen. Once a citizen has a certificate, a smart card, and a national ID, she can sign contracts, take out mortgages, and get loans without having to handle any papers physically. This serves an important security function for the government, but also benefits banks, utility companies, ticket merchants, and others. Any business that deals with paperwork, commerce, or identification can process more transactions quickly and securely by leveraging the active infrastructure of the national

ID. Over time, the increased commerce should more than pay for the initial security costs.

Of course, national ID programs are not the only examples of active security. Any system that focuses on making life better for all legitimate users can be set up to yield positive economic returns.

Two Basic Questions

For any transaction to be secure, whether it is physical – getting on an airplane, for example – or logical – opening an email – two questions must be answered. First: Are you who you say you are? Proving your identity and establishing who you are is authentication. Second: Are you authorized to do what you are trying to do right now? That is validation.

The first question, proving your identity for authentication, has been the focus of most software security technology over the past ten years. There are many different ways to achieve authentication, including biometrics, certificates, and secure IDs. An important aspect of authentication is that it can be done while disconnected. That means you can establish who you are directly with whatever you are trying to access without requiring some kind of centralized, trusted authority to confirm your identity.

If you have a smart card with your fingerprint on it, for example, you can swipe that card and put your finger on a sensor. The sensor will match your fingerprint with the fingerprint template that is already on your card and do a local match. No other transfer of information is required. With these techniques it is fairly easy to prove your identity with a very high degree of confidence. However, that's only half of the problem.

The issue of validation – sometimes also called authorization or access control – is more difficult and has been mostly ignored until now. The only way to perform actual validation in the past was to make a secure connection to a centralized server for each action. A person first had to have his identity confirmed. Then a connection was made to a master authority that had a list of everything that person was authorized to do. This scheme works well with a small number of users, but is totally impractical with hundreds of thousands or millions of people.

Think about credit card authorizations. Every time you buy something with a credit card, the merchant contacts a central authority to check your balance and authorize payment. This is a typical example of a validation system that is centrally connected, and the only example of a centrally connected system that works for millions of people. However, let's look at the characteristics of the credit card authorization system:

- The banks and credit card associations have already invested billions of dollars in the dedicated infrastructure.
- Each transaction is slow – it typically takes 30 seconds or more to run your card.
- Each transaction is expensive – the merchant usually pays 25 cents and 2 percent of the transaction.
- The system is not very reliable or secure – connections are sometimes unavailable, and fraud is common.

So, if you're developing a system and don't mind spending a few billion dollars, waiting a minute, paying for every transaction, and having poor overall security, then you can use a centrally connected model for validation. This is barely tolerable every time you pay for something with your credit card, but *completely unacceptable* if you had to do it every time you go through a door, read an email, or sign a document.

My company does scalable, distributed validation. With CoreStreet's Real Time Credentials ™ technology, you can make sure that any person, anywhere in the world, can prove they are allowed to be doing whatever they are trying to do – right now, without any transactional expense, even if they don't have any network connectivity. Combined with standard authentication technologies, this is a small but vital component of any large-scale active security system.

A Platform to Build On

Active security involves having a unified credentialing infrastructure. Unified credentialing deals with two issues: proving who you are and proving you are allowed to do what you are trying to do. The purpose of the infrastructure is to create the framework on which you will put revenue-generating applications. The infrastructure is an investment, but once you have it, you can put applications on it that people want to use – applications that are not only about security, but that also produce positive economic returns. The infrastructure makes the applications possible. Moreover, because the infrastructure is based on active security, rather than restrictive security, it enables applications that will pay for the infrastructure itself.

The applications that sit on top of the infrastructure enable more efficient types of legitimate transactions. An example of the benefits of using an authentication and validation infrastructure is secure mail. It allows you to send email that is secure, that no one except you and the recipient can read. More importantly, secure email allows you to validate who you are and who the recipient is. If you are in the military and you receive an email with orders from a general, you can be certain the email, in fact, came from the general and that the sender still has full rights to issue those orders. Once you have the authentication and authorization infrastructure, that type of secure, validated transaction is invaluable.

Secure Email	Secure IM	Single Sign-On	Doc Signing	Trusted Traveler	...	Physical Access	Applications

Validation (Are you supposed to be doing what you're trying to do, right now?)	Infrastructure Platform
Authentication (Are you who you say you are?)	

Another example is instant messaging, which is a problem, both in corporations and in the government. People love instant messaging, but currently IM is not secure; that is, it is easy to intercept and does not leave a sufficient audit trail. A more important concern is that there is no guarantee the people you are messaging with are who they say they are or that either side of the conversation is allowed to participate in the discussion.

If you could use IM securely, you could have extremely efficient communication. Many organizations forbid using IM in official business because it lacks security. Enabling secure IM is an example of an active security approach. It allows people to do their work better, faster, and more easily than they could before.

Besides secure communication and collaborations, applications that can leverage an active security infrastructure include single sign-on, document signing, trusted traveler credentials, digital content distribution, and electronic commerce.

An example of a physical security application is a door lock. CoreStreet is working with the world's largest manufacturer of locks, a company called Assa Abloy – which owns HID Corp., Timelox, Sargent Manufacturing, and Yale locks, among many other companies – to embed CoreStreet's Real Time Credential Technology into Assa Abloy's door locks for access control validation.

Our goal is to put smart locks everywhere, even in places that do not have connectivity. The locks will make sure that only people who are supposed to be allowed in actually get in. For example, a telecom field service employee is supposed to maintain a telephone company switch box in each neighborhood. With this new security technology, the service rep can open the boxes without having to remember passwords and without having to make copies of keys. And, of course, unauthorized employees will not be admitted.

The Best Victory

Recent war, terrorist attacks and well-publicized computer crime have forced us to reconsider how every transaction is processed. Such attacks are intended to inflict loss of life and significant economic damage in two forms: direct damage from loss and destruction and long-term damage to the system.

The effect the September 11th attacks had on the airline industry provide an illustration. There was the initial physical damage of that day: four airplanes were destroyed, and several hundred people on the airplanes were killed. The direct economic impact was further felt when all air traffic was suspended for a week, and the airlines lost billions of dollars. An even bigger economic impact to the industry was caused by people choosing to avoid flying. Moreover, the airlines and the government have had to spend more on security than ever before. The new level of spending may well be more than the system can support. That is the long-term damage – a real challenge.

A pernicious, long-term danger of the recent attacks is the harm that continues to be inflicted on our economy as we struggle to implement burdensome and expensive security responses. We have been prodded to change many of our security procedures and technologies. Our best victory will come if these changes result in active security, which will succeed in making us not only safer, but economically stronger as well.

CoreStreet, Ltd. is a company that enables organizations to grant or revoke an individual's access rights to buildings, computer networks, laptops, and wireless devices immediately in real time.

As president, Phil Libin is responsible for leading CoreStreet's engineering efforts and product design strategy. He is a skilled technologist with both hands-on development and managerial experience.

Before CoreStreet, Mr. Libin was founder and CEO of Engine 5, a leading Boston-based enterprise software development company acquired by Vignette Corporation in early 2000. At Vignette, Mr. Libin served as principal architect and chief technologist for applications. Prior to Engine 5, he led a number of software consulting and technology projects at ATG, Xchange, and EF.

Learning to Adopt Security Proactively

Firas Bushnaq
eEye Digital Security
Co-CEO, CTO

Conversion from Reactive to Proactive

As a relatively new industry, network security has long been trying to overcome the misconception that security is a reactive mechanism. Security is not reactive; it is not a burglar alarm for the network to let you know when someone is trying to break in. Security is about implementing policies and systems that will make sure a company is secure from the break-in to begin with.

Over the past two years, however, there has been an increased awareness in the marketplace of the importance of being proactive regarding security issues. For instance, instead of waiting for an attack to happen and trusting that a firewall is configured correctly, we seek to deliver protection against outside threats by constantly assessing a company's risk level. Companies are becoming more open to that proactive approach to security. This proactive approach is a conversion that is happening over time.

The Life Cycle of Network Security

The network security industry and the current product offerings within the security market space all run on one basic idea: protection from attack. The mechanisms for protecting against attack can be broken into three different segments. The first segment is pre-attack security; the second is during-attack; and post-attack is the third segment. These three make up the timeline of the security life cycle.

Pre-attack security involves a company's security policies, proactive risk assessment procedures, education and training, "hardening" processes, and anything else used to improve the company's network's exposure to external threats. Vulnerability assessment, policy auditing, and patch management software packages have a home in this segment. The concept behind pre-attack security is simple: Find the weaknesses, assess their risk, and develop a plan to remedy them.

The during-attack segment is the reactive part of security, where a company is reacting to an ongoing threat that might be attacking its network. This attack may be in the form of a hacker trying to break in or a worm that is attacking the machines on its network. Software and hardware deployments for this segment usually include network and application firewalls, as well as intrusion detection and intrusion prevention systems.

The post-attack segment, which is the forensic aspect of security, audits what happened during the attack. This segment has two roles. One role is assessing the damage and discovering what has been compromised and how from an actual attack. The other role has to do with assessing the successful and unsuccessful attacks against the network and using that information to better protect it in the future. While intrusion detection systems provide some forensic data, traditionally network traffic analyzers are used to record and play back the exact details of the attack.

The Philosophy Behind eEye's Products

eEye Digital Security is a software company. We focus on understanding our clients' real-world needs, and then articulating those needs into products and solutions that can help their security issues. With our vision of the security life cycle in mind, we have worked to develop a product line that provides protection pre-attack (vulnerability assessment), during attack (intrusion prevention), and post-attack (network traffic analysis). By working toward the goal of a complete security software offering from day one, we are now able to focus on creating new products and trend-setting features for the existing products.

Because security is evolving from a departmental responsibility into a corporate concern, eEye's products have evolved from segregated, departmental applications into enterprise-wide solutions. Your security is as good as your weakest link in the chain. There is no certain segment in which a company is more vulnerable than others, and once you open up one area, you end up exposing the rest of your internal systems. Thus, the security products being used across different departments, and even different office locations, should be able to communicate and provide a complete picture of the enterprise's security.

The three business solutions for covering pre-attack, during-attack, and post-attack security combine to provide comprehensive protection and information for continuous improvements of a company's network security. These solutions are:

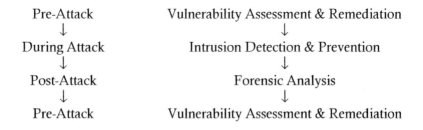

Pre-Attack   Vulnerability Assessment & Remediation
↓   ↓
During Attack   Intrusion Detection & Prevention
↓   ↓
Post-Attack   Forensic Analysis
↓   ↓
Pre-Attack   Vulnerability Assessment & Remediation

Enabling the Security Process

The solutions above only provide the tools and the means for a secure organization, but for security to happen, the tools must be adopted into a well-planned set of processes, and they must be carried out consistently and diligently. Generally speaking, practices for creating a security network involve education, assessment, follow-up, enforcement, and so on. At a higher level, we see a continuous cycle of best-practice security methods, one for each segment of the security life cycle:

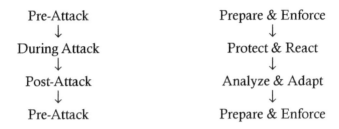

The solutions enabled by our products, when correctly implemented and adopted into a corporation's security practices, enable this proactive cycle.

Cutting-edge technology has been and will continue to be the only driving factor in pushing our company forward. Starting now, our focus is on the convergence of pre-attack, during-attack, and post-attack technologies. Our current solution offerings, as well as similar offerings in our industry, can be implemented together to provide adequate security coverage; however, they are still separate solutions. Our vision is to integrate the technologies in all three phases of the security life cycle so that they can share information to work together as one encompassing solution. The better our products work to enable the cycle of security processes, the more likely the security processes will succeed for an organization.

Expanding into the Enterprise

As mentioned earlier, network security is fast becoming a concern of the corporation as a whole, and less a departmental function. Security products first developed for the single administrator have undergone changes, and at eEye those changes have been successful as a result of client feedback and a flexible, componentized software architecture.

There are two factors involved in creating a successful product. One factor is its ease of use and implementation. The second factor is the actual effectiveness of the product overall. Feedback from our customers provides the most essential information for making products that are effective and usable but without disrupting business. At the same time, we also seek to create products that complement pre-existing security implementations and business processes.

Our company generates profits by constantly being innovative about the implementation of security. When adopting a new security process, companies must undergo changes to current practices, which may cause disruptions and inconvenience. The more inconvenient a security implementation is, the more frustration there is, and the greater chance that the acceptance of the security process will fail. We remain in very close contact with our clients, and we listen to the large corporations and small businesses that are using our software and make every effort to understand their concerns. We then try to design our products to eliminate their difficulties.

From a departmental level all the way up to the enterprise level, implementation and scalability are also very important factors in the success of a product. When used at the enterprise level, implementation can sometimes be done on a large scale, which can be extremely hard to manage. That is an important consideration for us when we build features for the larger enterprise into our products.

To address the problems typically seen in large-scale software deployments, we have adopted a componentized, modular approach to our technology architecture. This allows us to be more nimble and reactive in our product development and simplifies product support. For the benefit of our customers, as well as ourselves, new features can be added and tested quickly, and installation is simplified. Our customers also appreciate that our products can be implemented alone or in any combination to provide the optimal solution for their needs, and that additional products can be added at any time with seamless integration.

Finally, we make sure that all enhancements to our product line are consistent and work logically with existing functionality. We are not in the business of custom development, so when a customer makes a request, we put effort into understanding how the request could benefit other customers and improve the product as a whole.

Security Research Helps Security Development

eEye is in the vulnerability business, so our focus is on understanding weak points and how they affect systems. We concentrate on how to detect vulnerabilities because knowledge is the first layer in security.

Then we look at how we can eliminate or neutralize a threat, and then automate the elimination techniques so they become part of the normal internal process. We focus our research and development around complementing these objectives and encourage the engineers responsible to concentrate on achieving those goals.

The Stages of Vulnerability

Weaknesses in operating systems and software packages are a major concern of security-minded organizations. These vulnerabilities provide open doors to a company's network and therefore must be identified and patched. A software vulnerability undergoes a basic series of events:

1. A software product containing the vulnerability is released to the public.
2. The vulnerability is discovered and reported to the software manufacturer.
3. A patch for the vulnerability is released.
4. At the same time the patch is released, the vulnerability is disclosed to the public.
5. Programs that exploit the vulnerability on un-patched machines may be created and potentially shared.
6. Worms and scripts that automate exploit code are created.

The length of time a vulnerability sits undiscovered in a piece of software may be infinite. After its discovery, the vendor may need many weeks to develop a solid fix for the weakness. The length of time between stage one and stage three is what we call Pre-Day Zero Window of Vulnerability or Pre-Disclosure Window of Vulnerability. Risk during this time is low but significant. If an exploit for the vulnerability is developed before the patch is released, in many cases machines would sit defenseless. The time between when a patch is released and when all systems within an organization are secured is what we term Post-Day Zero Window of Vulnerability or Post-Disclosure Window of Vulnerability. This window can be minimized by the proactive efforts of the company, but the chance of exploits and worms is potentially greater, increasing risk.

Securing the Cause, Not the Result

We started this business with vulnerability assessment utilities, risk-assessment utilities, and remediation and task management tools. Using this approach, we protect companies against their vulnerability to the outside world. Focusing on the vulnerability prevents us from having to build a new solution every time a different exploit or worm appears for a vulnerability that has existed for months. Rather, our key approach is to be prepared from the time the vulnerability is discovered and to implement the solutions that protect our clients from the outset.

For example, intrusion detection systems rely on a set of signatures that identify known attacks. For one vulnerability, there may be several attacks, and for each new attack method identified, a new signature must be created, tested, and pushed out to customers. This is why we recommend an ongoing vulnerability assessment process to eliminate the threat of attack on critical machines, and on as many machines as possible at the outset of the Post- Day Zero Window of Vulnerability. In conjunction with vulnerability assessment, however, intrusion detection systems provide valuable support to secure a network until 100 percent of machines are patched.

From Research Knowledge to Software Innovation

We do a lot of research with regard to how vulnerabilities will be exposed within the networks and how hackers might take advantage of these vulnerabilities in the future. We challenge our groups to look at both sides of the industry – from the side of a person trying to break into the network and from the side of the person protecting the network. We then use that information to determine what security issues will be coming next and develop products accordingly.

Understanding vulnerabilities helps us build innovation into our products to protect our customers while the vulnerability is still unknown and exposed – the Pre-Day Zero Window of Vulnerability. The more we can learn about how and why vulnerabilities exist, the more we understand that almost every weakness falls into one of a handful of classes of behavior. With this knowledge, we can create smarter technology for our existing product line, and better secure our customers.

Keeping an Edge

At eEye, our biggest marketing success comes from sharing our security research with the industry. We seek to help the security community communicate in order to better understand our results, which helps organizations understand the value of security products. This sharing of information gives us great exposure and helps us promote the fact that we have one of the brightest and most innovative groups of security professionals in the business.

We also maintain our edge by being very active in the industry. We go to many industry presentations and conferences and are very active in presenting at conferences. We interact with researchers from other companies and try to share enough knowledge to be able to gain more equity in the space. We also keep ahead by investing in our own employees. We provide them with the ability to engage in continued education so they can thrive in this marketplace.

Advice for Companies Thinking About Security

A company uses security to protect its assets. As long as a company has digital assets that need to be protected, security is a concern. If a company is saving digital information, it is in its best interest to keep the information in a safe place, away from prying eyes and competitors' eyes, and to be able to reassure its own clients that their privacy is preserved. Every organization has a responsibility on behalf of its clients to do everything possible to make sure protection is in place.

The security – or lack of security – of operating systems has brought attention to the security marketplace. In the past, security was more or less limited to the IT level. However, current exposure has brought about new companies and new products, and now consumers will continue to be more cautious about the security of their deliverables. In addition, the Internet itself has generated exposure to security threats across the world. Ways to protect against such threats have accelerated the security industry. At the same time, the conversion from a reactive to a proactive approach to security has helped the industry become better at dealing with security, and still more will be done in that area.

Make Security Part of Every Business Practice

Security is a means of protecting your organization from internal and external threats to your assets or anything of value. It involves a set of processes, implementations, tools, utilities, and systems that are set in place to protect against these threats. We incorporate security in almost every business process for our clients. Implementing security within the different processes and making it the forefront of implementation is critical.

The majority of companies are vulnerable to security breaches. The extent of their vulnerability depends on how much a business has invested in its security. My advice to businesses is to have reactive systems in place, and also to be proactive, which is the most important piece. Businesses must think of security as part of the normal way of doing things. At the same time, they should continue thinking about potential risks and how to eliminate those risks before the actual threat occurs.

As for companies, it is critical to think about security every step of the way. They should not rely on one individual mechanism for security. Rather, they should think of security as multiple steps to go through to protect their assets. As a rule, a company should think about every piece of data that resides on the network, including who has access to that data and where the data moves to and from, starting with the most critical elements.

Make Security a Never-Ending Practice

Security should not be viewed as an end or an attainable goal. Instead, security should become an inherent part of every person and every process.

For executives especially, a common misconception – or misdirection – is that security should be treated as a single item on an agenda that needs to be implemented. Executives take an approach that says, "We need our end-of-quarter result systems to be in place; we need our accounting and ERP systems to be in place; and, oh yeah, we need our security systems to be improved." While security spending may decrease over time once the proper tools are implemented, security as a whole should never leave the minds of those in charge.

Empower Security-Minded People

The biggest challenge today in the security business is awareness at the executive level of the importance of security. The way we have been growing over the last five years is by creating software at the departmental level. The IT administrator, the security administrator, and the person who deals with security on a day-to-day basis understand the value of our tools, the products we implement, the processes we recommend, and the remediation management we push into the marketplace.

By contrast, executives have only just begun to appreciate the importance of implementing security inside the corporate network and establishing a budget for it. Executives are the ones who will open up the potential for the exploration of security projects that cover all the internal processes of normal business activities. Executives also have an interest in purchasing security products that cause the least disruption.

The most critical thing to remember is that although security may be a distraction, it should never be an obstacle. The value of security far outweighs most inconvenience, and any employee who causes a slight disruption for the sake of security should be rewarded, even though they are often criticized.

Understand Your Limits

A company should also implement solutions that are the most beneficial to its particular business. There are many different security options to learn about. A company should learn everything it can about the available solutions so it can find the solution that works best for its needs. In addition, companies should invest in the top safeguards and should know how to react in case of a threat.

In this growing industry, it is easy to get overwhelmed by the mass of products, services, and advice. Companies should keep in mind that security is not all-or-nothing, and that a goal of being 100 percent secure is almost unobtainable. Implementing tools and process to protect the majority of assets, starting with the most critical assets, is enough for most organizations, no matter what people selling to you try to get you to think. For an organization just starting to implement security practices, sticking to a rollout plan and making sure each piece is accepted and functioning before starting the next ensures that the security processes will remain in place for years to come.

The Future

Security is a state of mind, so political issues, government issues, terrorism, and viruses all have an affect on this industry. No one has control over security threats, so security has to be very adaptive.

Moreover, the economy does not greatly affect the security industry. In the days of prosperity, when the economy is moving up and everything is growing, companies want to protect their assets. When the economy is down, and all a company has left is its assets to protect, safeguarding them is likewise important.

Getting Over the Reaction Period

One of the biggest reasons awareness has increased over the last two years is that the threat has become indiscriminate. Before two to three years ago, a threat would come from a hacker, a person who was maliciously trying to break into your company. Most executives have the mindset of, "I do not have enemies out there, and there is no reason for somebody to maliciously and illegally try to steal something from me. So I am not too concerned about that potential threat." However, over the past few years, worms and some of the viruses that have attacked are indiscriminate. They disrupt businesses everywhere, causing downtime in factories and in departments that were never worried about threats. Today things have changed, and the level of awareness and concern is greater.

This recent awareness forced companies into a sudden reactive mode to security. There was a rush of forced purchases to try to make up for lost time. Thousands of security companies came up with hundreds of ways to solve a few distinct problems. This overabundance of solutions and advice causes many problems for organizations because they do not have a standardized way of measuring how secure (or insecure) they are, or when they are secure enough. Security is essentially measured by how much a company has invested in it, instead of how secure they truly are. Vendors have tried to overcome this problem by developing policies and measurements for risk and exposure, but again the policies are not widely adopted, and the metrics are not standardized.

As the security industry matures, the industries around it will also begin to mature in the way they calculate risk and adherence. The beginnings of this can already be seen, especially in the government sectors. For example, the SANS Top 20 policy has been widely adopted, as has the ISO-17799 Standard. By using these standards, companies can easily determine specific areas for improvement and can have a greater sense of their overall security posture.

For the security industry to retain credibility, for security data to have useful meaning, and for security companies to make advances in common directions, security policy will need to be created by organizations outside of the industry. Once policies create a set of "industry standards" for security, security vendors will be able to focus on products to help companies achieve those standards.

Security at the Source

The future of security also lies in the hands of software makers. As network, hardware, and software security problems make headlines, it also focuses attention on the source of the problem. The attention needs to be positively absorbed by every person involved in computer science to make an impact on the next generation of products. This movement toward more security products will be driven by security research.

As companies work to understand the reasons behind vulnerabilities, the research will need to filter down to educational channels to be standard inclusion in course materials and curriculum. Research will also foster the production of software development tools, such as software auditing utilities and intelligent compilers that promote secure coding techniques.

There is no silver bullet or magic recipe for security vendors to follow for generating revenue. Revenue is only a reflection of the value a company brings to the client. It is all about understanding the client's needs and satisfying those needs. That is the only way to generate revenue and to be successful. Complementing that are good vision and forward thinking, which will lead to future revenues.

Firas Bushnaq co-founded eEye Digital Security in 1998 and serves as the company's co-CEO and CTO. Mr. Bushnaq started eEye Digital Security after the successful establishment of three previous software development companies: eCompany, WinWare, and Avataro.

For the last five years Mr. Bushnaq has used his knowledge of software development, architecture, and design to develop the technical foundation for eEye's security product offerings. He has also applied his entrepreneurial expertise and unique corporate insight to foster the growth of a well respected and successful security software organization.

Mr. Bushnaq has a BS in computer science and a master's degree in software engineering.

Solving Problems for Customers: The Path to Continued Growth

Joseph W. Pollard
Permeo Technologies, Inc.
CEO

Basics of the Security Industry

Security today is different from security in the past. Much of the security we may have thought was adequate five or six years ago was actually associated with user name and passwords and the users of the network. We wanted to make sure we gave access appropriately to those users on the network. It was mostly thought of as our employees on a local area net or wide area net getting access to resources on the network. Today, with partners and contractors, whether supply chain or support chain management, and with mobility coming into play and the Internet proliferation, security means not only providing the resources to the appropriate people – employees, contractors, and partners – but it also means trying to protect those assets from unauthorized use or exploitation.

I am the chief executive officer of Permeo. Our company is focused on helping all types of customers – banking, pharmaceutical, manufacturing, high-tech – extend their applications to employees, partners, and contractors. These applications could have been written ten years ago; today they could be Web-based, or Web services-based in the future. We try to make sure we help those companies get the applications securely through the firewall across any kind of network, whether it's wire line or wireless, to the appropriate users.

Vision for the Industry

Migration will take place in three different areas.

There will be increased amounts of security in just the hardware equipment we buy. Much of the software capability we have layered on in the past will be implemented into hardware for performance and manageability.

The second thing we will see is a continued emphasis on making sure the infrastructure that is in place within both large and small companies is protected through the use of layered software similar to the type we develop at Permeo. That is to enhance or to support their business units or business process deployment within the company. You want to make an "in-office" experience available to everybody, irrespective of location.

The third area is increased and continued focus on the attacks and the people who are trying to negate through security attacks the efforts of businesses to deploy applications to users in support of their business.

Drawing Customers

The reason people buy Permeo is the maturity of the software and its ability to provide secure access for any kind of application over any kind of network through any firewall. For instance, there are applications that are legacy based, two-tiered, and are just as appropriate today for use as they were five or ten years ago when they were put into production for managing businesses. Today companies are deploying Web and Web services and will want to use the same tools for managing secure access to all of the various types of applications.

The main value we bring to market and the reason people come to us is that they have a unique network environment, with a need to provide access to applications for their users securely and make sure that only people authorized are permitted access. We help them extend those applications to the appropriate employees, contractors, and partners.

Vulnerability in Customers

Most of the vulnerability comes from use of applications specifically like email. The speed at which some of these viruses have traveled and proliferated worldwide, like the worm virus, is amazing. Most of the speed happened because the viruses were tied to application usage and data flow. With the proper security mechanisms in place, users are not that vulnerable.

In security there is no silver bullet. There is not one specific thing that you can do to protect the applications on the network. Using multiple security products will create a defense-in-depth environment, meaning there are multiple security products that have to function at each layer of the OSI (open system interconnection) model to make sure people are truly safe in what they are trying to accomplish.

Permeo secures specific assets, such as applications on servers, so you have multiple layers of security, checking data coming in at layers two and three of the OSI model, for example, or as we do, application security and filtering at layers four through seven.

There are applications you try to deploy to external users. To deploy those through the firewall, in many instances, you have to open ports. As long as those ports are open, traffic can come in just like traffic going out. One of the single points of security vulnerability is open ports on the firewall. Through dynamic port management, we allow data to flow in and out and then during the wait stage between the data flows, we shut those ports down.

The first thing someone should get for enhanced security is a firewall. These can be deployed either as a personal firewall on your computing device or on the network for a group of users. That would be the first recommendation. The second recommendation is antivirus, making sure you are up-to-date with the latest releases from the antivirus companies.

Building a Plan for Security

The first thing we talk about when building a security plan is what applications or what part of the network are you trying to extend to business users, contractors, and partners. CEOs are often forced to make a decision: The more applications you extend out from within the corporation, the higher risk you put on the network itself. If you lower your risk, you forgo the option of extending out these applications. Most of our customers are large companies; they could be multi-continent and certainly are multinational. Not being able to extend the application infrastructure they have deployed in one area to another simply because of security risk is unacceptable to them. That is the problem we try to focus on.

The step is try to figure out the priorities for getting applications to the users. Are they internal users accessing internal servers that perhaps run CRM or an ERP application, or are they Web-based, or in the future will they be Web services? What is the business trying to solve by extending those applications out?

The third thing we look at is the infrastructure of the network. Is it a wide area net with a combination of wireless and wire line computing devices? How mobile are the employees, partners, and contractors? Getting the general infrastructure of the network topology is something we need to understand so we will have the ability to deploy applications to the appropriate type of user.

And finally, we take the data from the due diligence process that has been given to us and present it back to the client for both implementing the Permeo products and devising a plan for how that would integrate with other security products, which will give them a much tighter and more secure environment.

Staying Ahead of Hackers

As you can imagine, there are a number of steps we take to stay ahead of hackers. One is that we follow many industry institutions that make it their focus to understand where the attacks can come from and what applications are the most vulnerable, and in the case of Microsoft products, make sure that we understand which are the latest patches and what should be available out there.

The next step is to take a look at the architecture of a network and make sure it complies with the simple building blocks we think should be in place – firewalls, antivirus, intrusion detection or protection – and then integrate our product in this environment, as well. As new technology is introduced, much of the job comes down to looking for points of vulnerability that could be exploited by people who have a malicious intent. You cannot always be in front of them. Implementing the basics of security solutions will stop some of the malicious intent of hackers, but the job never ends.

Biggest Securities Issues

The biggest security issues in the future will be the wireless networks. There are hot spots popping up by the thousands and potentially the tens of thousands around the world. The kinds of applications that will be used on those – whether you are in a kiosk or have a handheld device that is accessing applications across some type of a wireless network – are all points of entry to an IP network somewhere. We try to make sure our customers understand a single entry point, if unprotected, is like a pipe coming into your network. It gives anyone who enters access to your company applications and assets.

For security in companies that have assets you can see, you use the same philosophy of defense in depth, or layered security. A company that has physical assets that could be removed would put up a fence around the perimeter, which is perhaps analogous to a firewall. You use camera technology to survey certain areas that contain assets. You have security guards. Whether the assets are electronic or physical, you still have the same approach, and that is what we have to help people understand. The approach is similar, but the tools to accomplish the end are different.

Building Trust with the Client

Trust can be gained only over time and with experience. When we do not have the benefit of an existing relationship, we try to show customers how we have solved similar problems with similar types of companies. This is just the maturation process of an industry. People have problems, and although they don't know the solutions, they understand that they are probably not unique. We show them similar types of companies with similar problems and how we resolved those.

We have many reference-type engagements at this point. Customers want to talk to other customers and see how satisfied they are. Once you have solved one problem or taken a specific problem and that succeeds, the level of comfort with our company within the customer base increases, and customers give us other problems to solve. You don't have instant credibility – it has to go through a maturation process.

For a more secure environment, you have to be able to know and track where your assets are and be able to identify whether they are handheld, desktops devices, servers, and over what types of communication lines coming in and out of a company. It forces you to put other mechanisms in place to track turning up assets and turning off assets, and that process also allows you to look at utilization to decide whether capital purchases may be appropriate. In some cases, because asset tracking has been somewhat lax in the past, it allows you to go out and reclaim unused assets or make sure they are used to their full capacity before more capital is expended. That is just one of the byproducts of knowing what your architecture is, the implementation and the assets on it, and how you are protected.

Customers are ready for a security solution at various states, depending on the size of the company. The minute you put computer and software assets into your corporation, you are ready for your first implementation of a security product, even if it's just an antivirus on your personal machine in a small company.

A large company will go through different phases to roll out security. You have to start with the perimeter and then work backward to hard assets as servers and the applications that run on those servers. From day one, everybody is ready for some type of security. The use of applications and the size of the company will also tell you the types of security investments that will be needed.

A smaller company may be interested in just remote access management for email. They may have 50 people outside their headquarters – a field sales force, for example, engaging with customers, in different geographic locations from the mail server. In that case, those people would be ready for some type of an SSL (secure sockets layer), VPN (virtual private network) remote access solution, which Permeo makes available.

In another situation where you have a global company that has offices in North America, on the European continent, and in the Asia Pacific area, then you have not only a geographic disparity, but you also have users from different parts of the world trying to get into different parts of the network 24 hours a day seven days a week. We must be able to offer products that can be implemented in North America, Europe, and Asia, and to facilitate extending the application environment to someone who may be on the other side of the world. You then look at the geographical area to decide what parts of the network they are accessing. Is it wireless? Is it wire line? What kinds of protocols are they using? You have to have three- and four-dimensional perspectives of the network.

Even though all these companies are ready for security, the solutions they would implement would be substantially different.

Making a Product Successful

Two things specifically make a product successful. One is that it fills the gap or eliminates the unsecured scenario you are trying to solve. The second is that it not only solves that problem, but it can be extended, or it can grow, over time to allow you to solve other problems.

If you are trying to secure email, for instance, that may be just the first of many, many applications. The Global 1000 and the world's largest companies have hundreds of applications. In one case it could be just starting with electronic mail, and the next could be CRM or ERP, or some of the large computer-intensive applications. It could be Web services being layered on top of those applications and others in the future. To be successful, the software not only has to solve a problem today, but it also has to be dynamic, and you have to be able to continue to add features to the product that allow you to deal with more and more of the same types of problems, with different applications on different networks.

R&D and Marketing for a New Product

When deciding that we want to add functionality to a product, we spend a lot of time talking with our customers, partners, and value-added resellers, making sure we understand what kinds of problems our customers are asking us to solve today. Then, based on the roadmap of deliverables within the business, we look at what kinds of problems they anticipate that we'll need to solve tomorrow.

We gather the data and run it through our analysis here in the company, deciding which options fit our business and which ones don't. We then feed the marketing requirements documents and product requirements documents, called MRDs and PRDs, to Engineering that state specifically what functional enhancements are required in the code. Engineering will review those, and any issues will be resolved between Engineering and Marketing.

We divide the Engineering tasks by capability within the Engineering organization. Some people may work on the user interface, and some people may work on the installability of that portion, and others may work on the specific functional enhancement. All of those pieces are brought together, and then we start our internal testing and QA process, followed by the external process, which is the beta testing.

We then think about the best ways to get our message to the intended audience. For instance, we develop white papers about the specific subject and put those on our Internet site so that people can download them for review. We do targeted mailings. We attend industry forums, where we can present to audiences of various sizes. All of these things are building blocks around one specific issue: How do we get the message and the information to the intended users and the audiences of the companies we sell to?

When we decide to release new functionality or enhancements to the product, we actually will go back to existing customers that already own the product and talk to them about the new feature sets and enhancements. It works across the product lines, and the message is always the same: Here is another enhancement we think will improve the security in your environment.

Biggest Marketing Misconceptions

Often executives think that once they have completed the implementation of a security product, they are finished. In fact, they are not – that was just the beginning. Also, when executives view products, they have to look at them from a layered approach. Different products solve different problems, depending on the configuration of the network and the utilization of the network. You can't implement one product as a silver bullet to solve all of your security issues.

The most important thing to understand about security is that you are never finished. The minute you think you are, someone is out there trying to exploit and take advantage of the vulnerabilities that are set up because of network growth or new products.

Another challenge we face is that security is so ubiquitous that it takes customers a while to get through information to understand what is best for them. When you think about networks, you can imagine all the different kinds of equipment, the software, and the geographically dispersed implementations. People understand *network* is a single word with many meanings. The same thing applies to security. When you say security, people tend to think antivirus or perhaps firewall. There are many, many different types of security products out there, and just being able to evaluate them and figure out whether they will solve your problem is a challenge for customers.

Continuing Growth and Revenue

To be sure the company continues to grow and generate profits, it comes down to constantly focusing on the problems that our customers are trying to solve today and then looking for the windows of opportunity six months, 12 months, and 18 months into the future. I like to think of things in six-month increments, in which we have engineering efforts developing enhancements to the product that will then be released within the six-month window. Then, within 12 months, we go back to customers and say, "Here is what we are working on for the next release, and here are some tradeoffs."

We want their opinions. We talk to partners about that and to customers and our resellers and ask them to help us prioritize the things they want us to do now, and then 18 months out, we look at things we could do. We try to get feedback from customers that we are on the right track and that those are the problems they see surrounding deployment of their software assets in their companies. We want to see if we are addressing the right areas and solving problems they believe could occur, based on their growth or their rollout of specific applications or assets.

This effort all goes back to one thing: solving problems for customers. As long as we keep solving problems, we can continue to grow.

In 2002 and the first quarter of 2003, there was so much uncertainty around the technology area that in some cases, companies said they needed products in the security space, but they were not spending money. They made the decision themselves that the priority of keeping cash and deploying it to other things was more important than potential loss because of theft of data or some type of attack within their network.

Each time theft or an attack occurs, a quantifiable dollar amount is associated with that fraudulent use or attack. Companies make the decision that security is something they just do not want to deal with at that moment because they are focused on cash flows and keeping the business moving in the right direction.

As economic conditions change and growth starts to occur, companies can take some of the money from that growth, depending on whether they are profitable or not, and redeploy that into other projects. It is our challenge to make sure that we are one of the projects that get the visibility at the management level for the investment. The economy definitely plays a role in the decision. The economy plays a role in determining the particular types of risks that are acceptable or not acceptable during phases of an economic expansion or contraction.

When our customers are in a position to enhance their business by extending applications to users in support of the business objectives, and we make them more secure and provide connectivity that will enhance their position in their markets or with their partners or customers, then they are more willing to do that. Once we see that, we know we have to engage and help them deploy those applications to the businesses that require them.

Looking to the Future

We constantly stress to our employees and our partners that this business changes every 90 days. If you feel like the business has not changed, just wait and it will. Every quarter we look up and see whether we are still focused on the right things and whether we have the right investments placed at the right areas that will allow us to solve our customers' problems and get a return on our investments – the opposite of stagnation. You just preach constantly that you are in a state of change so that no one ever believes the business will stagnate.

We are looking forward to our ability to continue working with the Global 1000, the biggest companies in the world, and expand our ability to help them with certain security issues around the application and the extension of those applications through firewalls and across networks.

The reason we are excited is that we think we are well positioned in that space and that for the next five years or so, we can see ourselves growing and continuing to provide application security products to our customers. We can see ourselves working with the largest companies in the world in many different areas. We are excited because we do not see an end to the need for what we have, and we know we can grow our company around that philosophy, be profitable, and give our investors a return on their investment.

To stay prepared for these opportunities, we are constantly reading articles written by subject matter experts from industry magazines and documents to stay on top of the latest information. We go to many forums and industry-offered engagements, where we can talk to other companies like ours, customers, and chief security officers, as well as the consultants that help our customers. If we constantly evaluate the information we get from each one of these opportunities, we get a good feel for what is happening in the industry. Each of us tends to have a focus for where we get our information, and assimilating it gives us a good feel for what is going on in the minds of our customers.

You Are Never Finished

The best piece of industry advice on security is that you are never finished. It's similar to the lessons we learned in the early days of quality assurance: The minute you think you are finished, you have been beaten. As networks grow, and as the need for collaboration among businesses and employees within the businesses grows, the constantly changing networks and the deployment of those networks will drive more and more issues around security. The best advice I can give anyone is that you are never finished with security implementations.

The advice I give executives is to pick the vendors that solve your problems and stay with them over time, so that you can build a business together. The more I am integrated within a company's security needs and the needs for deployment of applications, the more I understand about their business, and the more I can anticipate and bring functionality to them that will help them extend their business models. I hope that means increased growth and profits for them, as well.

In understanding our customers' needs, we need to understand their geographical distribution, deployment, and the kinds of applications they already have. Most of our customers do have some security products already. We look at what they already have, where the holes are, where they are trying to take the business, and what kinds of issues that will present in the future. This gives us the ability to do a summary of that specific business and make recommendations about not only what kinds of problems or challenges we can solve today, but where they will need us in the future and what kinds of solutions we can deliver to them, based on the roadmap.

Major Changes in the Security Industry

The focus in the security industry has gone from users in a local area net to users and their partners and people who are not specifically employed by a company, but who use the same network to communicate with each other to get the business done. What used to be just a concern about users within the company getting access to data that perhaps they should not have, now it means opening your network to companies for business reasons, when you have no control over the users and it's a risk you have to take. To promote that business arrangement, you have to focus on your own security and protection within.

Some security products are security by exclusion. In today's world, where collaboration is such an issue, we have to be able to have security by inclusion; that is a big change.

The major changes in the future will be more and more hardware that takes security into account. The application developers have been and will continue to be focused on what they need to do within their own applications, and companies like Permeo will continue to focus on an environment where the customer has many different vendors and many different suppliers, local area nets, wide area nets, geographically dispersed, and static and mobile work forces. Our real focus is to make sure we can help them in any type of environment.

Golden Rules of Security

The golden rule for us is to stay focused on the areas where we can truly provide solutions to customers to solve their problems, and to make sure we understand what they will require from us in the future so that we can be considered a partner as they continue to expand their business. We have to make sure we don't try to solve so many problems that we are ineffective in any one area. We have to focus on the things we can affect and let other companies focus on the areas that we don't. If there is an area no one focuses on, it's an opportunity that probably will get exploited through the start-up or venture capital world.

The worst thing you can do is define your footprint so large that you don't make an impact in any one area. We have to constantly test ourselves on that.

There is enormous opportunity in the security space. With such great opportunity can come confusion because of messaging and positioning by various companies. There are areas where confusion and maybe some chaos exists because of multiple messages or the same messages from many different companies. Again, it is that maturation process of the industry.

In projections made by consulting organizations, within the next three years, just the security products – perimeter-based security, such as firewalls, antivirus, intrusion detection, and application security – will be upwards of $25 billion worldwide.

Although security is different today from years ago, the concept of security truly has not changed in the last 25 years. It is a combination of addressing perimeter security and then securing the specific assets of a corporation that make for a successful and secure environment. Maybe 25 years ago those were physical assets you could see and potentially touch. Today it is electronic and intellectual property and things that companies hold dear to their existence. The original philosophy has not changed. You have to have layered security to protect those assets.

Joseph W. Pollard serves as chief executive officer of Permeo Technologies. He has more than 20 years of experience in the computer technology industry, working with various start-up companies.

Most recently, from February 1997 to May 2002, he held positions of executive vice president and vice president of worldwide sales for MetaSolv Software, a developer of Operational Support Systems (OSS) for the telecommunications industry. During his tenure, MetaSolv grew from $11 million in revenue to $132 million. From July 1992 to February 1997, Mr. Pollard held director-level sales positions for both international and North American geographies of Tivoli Systems, a provider of systems management software. Before Tivoli, he held various sales positions for Visual Information Technologies, Apollo Computers, and Digital Equipment Corporation.

Mr. Pollard graduated from the University of Memphis with a bachelor's degree in engineering technology.

Simplifying Security from the Inside Out

Thomas Noonan
Internet Security Systems
Co-Founder, President & CEO

Security From Generation to Generation

The security industry has become very complicated. I would encourage any executive – or any person dealing with security – to go back to the basics. Usually, the experts begin a security discussion with "speeds," "reads," and "feeds," technical terms that alienate the audience from the start. Protecting your business doesn't have to be complicated.

Security essentially has two functions: to keep the bad guys out and let the good guys in. A comprehensive security solution also has to be able to manage changing circumstances, for example, when an employee becomes disgruntled, or a "good guy" goes bad, or worse – becomes a network saboteur. Executives should consider every security investment from that perspective to avoid purchasing a security solution that cannot respond to all of these situations, as well as changing business needs.

The most effective way to safeguard your company today includes technologies that are now becoming available. The security model has been perimeter-centric since its inception. Going back to medieval times, people built walls around cities for protection. But now, in the age of e-commerce, this model no longer works. The advantages of being connected via the Internet are too great.

Companies have gained enormous efficiencies from automating customer, supplier, vendor, partner, and employee communications and transactions via the Web. The only downside to this development is that the Internet becomes a new avenue of attack, making critical business systems vulnerable. The natural reaction is to put up another wall to keep the bad guys out; in this case, a firewall. However, even today's firewall technology poses a huge dilemma: Firewalls can guarantee protection only by keeping everyone out, an option as certain to jeopardize a business as having no protection at all.

Companies continue to use firewalls, but they've adjusted them to let all kinds of Internet traffic through. Today firewalls must remain so porous that they can't properly protect the network. Firewalls can't determine whether it's truly a good guy coming in, and they have no way to keep the bad guys out short of blocking entire traffic streams. Not until recently did the security paradigm begin to shift, with companies changing from a perimeter-centric to a multi-layered approach.

Today companies have no perimeters. Everything is exposed to the outside, ushering in a new era of multi-layered security that involves different security solutions to address threats at every level of the network.

While this was progress, businesses quickly found that the multi-layered approach has drawbacks of a different sort. Putting multiple security devices on the network becomes a tangled web that is very expensive and complicated to manage. Companies have to apply a new security solution to address each new type of threat. There's intrusion detection and intrusion prevention to stop worms, hybrid threats, and malicious code; anti-virus for viruses; content filters to block unwanted Web content; and now anti-spam products to get rid of unsolicited emails. Acquiring, installing, managing, and updating all of these technologies place an enormous burden on companies.

Considering the complexity still associated with security, executives continue to seek a better solution. The security requirements set forth in various government regulations place more pressure on companies to ensure adequate protection. The first step is for executives to prioritize security measures. They must first take necessary steps to ensure protection that complies with government mandated standards, such as HIPPA and the California Senate Bill No. 1386, both of which strive to protect consumers' private data. The next step is to apply security measures the company needs to adopt based on business demands, and the last priority is for companies to adopt policies and technology that they are afraid not to adopt because of the undesirable consequences.

Most companies are coming to the conclusion that they need a flexible protection system that is capable of consistently detecting, analyzing, and maintaining the gentle balance of allowing business to function in an open atmosphere while blocking malicious intruders. The answer is a unified protection agent, based on sophisticated intrusion analysis and prevention technology, that is capable of automatically detecting and preventing every threat from causing damage to the network. Whether the danger is a policy threat, an activity-based threat, a content-based threat, or a threat associated with regulatory exposure – one technology is capable of blocking them all.

This next generation of security technology is revolutionizing the security industry. Instead of managing multiple layers of disparate systems, a single unified protection agent can effectively protect across the entire network, eliminating the need for single protection agents, such as the firewall and anti-virus software located throughout the system. The economics of this technology are significant. Most CEOs don't realize that 80 percent of their company's security budget is delegated to labor, while 20 percent is actually spent on protecting their corporate assets. The vast majority is spent on security personnel reacting to the inefficiencies caused by numerous incompatible systems. This concept of the universal protection agent finally solves the main obstacle in the security industry – the return on investment. Most C-level executives can recognize the cost effectiveness in adopting one technology to do the work of five single products.

Assessing Risks

For companies to be able to protect themselves against malicious threats, they must first be able to identify the risks facing their organizations. Risk is defined as the intersection of a vulnerability and a threat, which may be a simple concept in the physical world, but is much more difficult to understand electronically.

By connecting to the Internet, individuals link countries around the world, creating a new universe where people cannot only be invisible, but can also change their identities at will. Unlike the physical world, Internet users are not required to provide proof of identity to act. In essence, users have the ability to anonymously launch the equivalent of a weapon of mass destruction, creating havoc in the physical world without repercussion. Look at the Melissa virus, Code Red, Nimda, Sobig, Slammer, MS Blaster, and the Mafia Boy denial-of-service attacks. Anyone who happened to be connected to the Internet at that time who didn't have the necessary protection to block these attacks – whether they were directly targeted or not – was negatively affected. These attacks caused real losses in business productivity.

When asked about security precautions, security experts often hear, "Why would anyone want to break into my system?" Many businesses have the erroneous belief that if their system doesn't contain information that would somehow benefit a hacker, then they are not at risk. What many CEOs don't understand is that many hackers do not try to break into systems for their gain, but simply because they can.

Any business that depends on an information system, the data contained on that system, and the proper functioning of that system, to run the business is at risk. That business, therefore, has an obligation to its customers, suppliers, employees, and shareholders to protect that information with a reasonable level of confidence. In fact, recent Sarbanes-Oxley legislation requires that businesses protect anyone relying on their network. California Senate Bill No. 1386 also indicates that any company that suspects its security has been compromised must contact affected parties and notify the public – which can be very damaging to corporate reputation.

The Internet was originally designed to be open and easy to access, making the challenge of safeguarding it that much more difficult. The best solution is to quantify individual risk. Many models are beginning to evolve that will be useful in helping CEOs determine what technologies are best suited for their organizations and at what level to establish their corporate security budget.

While meeting with the National Infrastructure Advisory Council (NIAC) at the White House, a member CEO from a non-security company, commented that he recently starting asking questions about his company's security. He found out that his company is receiving between 180,000 and 200,000 attacks and break-in attempts each month. He was concerned about how much damage those attack attempts would do if they were successful in getting into his company's network. He agreed that the result would be devastating. It would challenge his entire business and disrupt any existing progress.

Staying Ahead

Security is a game of cat-and-mouse. The only way we have found to stay ahead is to use research to remain fully aware and embedded in the security industry. Internet Security Systems (ISS) spends about 18 percent of its revenue on research and development to provide dynamic security. A typical security company spends approximately 10 to 11 percent.

In our experience, the most effective method of research is to hire intelligent security veterans who are passionate about this business to consistently monitor the Internet for unusual activity. Our security experts also provide managed protection services in 40 countries around the world to protect our customers' networks. As new threats are discovered, they are automatically supplied to our research and development team, X-Force, which comprises about 220 engineers who are analyzing those threats and immediately updating our systems around the world to stop the emerging threat from affecting our customers.

This is a 7x24x365 operation. Research cannot be performed on a 9-to-5 basis, or with a small number of employees, to be effective. Recently, businesses have recognized the advantages of original research and are demanding that their security vendors have the ability to provide this intelligence. Security companies that are not collecting and analyzing information cannot ensure that their technology, and therefore their consumers' networks, are up-to-date and able to block emerging threats.

In staying ahead of malicious attacks, it is also important to recognize where potential threats may arise. As a CEO, I worry about three kinds of general threats.

The first, and often most dangerous, is the quiet insider threat, which is difficult to detect because these corporate insiders are already operating on the network, are typically trusted, and are aware of what types of security technology their companies are deploying. In most cases, insiders conspire with security experts outside their organization to take advantage of their knowledge to defraud or damage the company in some way. If the business that is targeted does not have a dynamic security system that monitors the actions inside the network, that business cannot even see an insider threat, much less stop it. When considering that many large enterprises continue to deploy only first-generation security technology, such as the perimeter firewall, you realize why the majority of businesses cannot protect themselves against an insider threat.

The second type of threat that keeps me up at night is a coordinated, deliberate attack against an industry or asset class. The idea of a deliberate cyber attack was discussed heavily following the terrorist attacks on September 11, 2001. Security experts have debated the likelihood that a similar attack could occur on the Internet that could have the same destructive effects in the physical world. A coordinated attack against the Internet systems that maintain the transportation, banking, or telecommunications system – all fundamental pieces of the infrastructure for all citizens – would be disruptive to our entire economy. For example, if an attack succeeded in taking down the telecommunications system, there would be no way to communicate, and therefore no way for business processes to continue. While the attack is focused on one particular target, the effects are not easily contained.

Many may feel that this is a national security concern that does not involve the business community. However, considering that businesses control 82 percent of the Internet where critical infrastructure is maintained, it is obvious that the business community plays a role and must participate with the public sector to protect national security.

The third threat many companies face is the lack of awareness and understanding at the CEO level of how exposed their businesses are, including the billions of dollars spent to automate their businesses. This weakness has nothing to do with an individual threat, but the idea that security has still not evolved from a technical concern to become a mainstream business issue. IT or security professionals can attempt to convince senior management that corporate security should be a priority, but CEOs typically do not appreciate this concern until it's too late and their system has been compromised by a malicious attack.

To prevent this from happening, organizations should consider employing a security company to perform a penetration test to determine how vulnerable their business is to the outside. Another option is to use vulnerability assessment technology to scan networks for weak spots. I have presented at 15 meetings within the past year in which a company's audit committee has learned their organization's specific areas of vulnerability and immediately made it a CEO, or even board-of-directors, issue.

With proof in hand, every CEO consistently asks, "How long have we been this way, and why didn't we know about it before?" Businesses have been so focused on developing their information infrastructure that they haven't taken the time to consider how dependent their business has become on their networks, and in turn how essential it is to protect them. CEOs aren't getting the message until problems arise to motivate preventive action. It is unfortunate that in a time when capital spending in the industry continues to decline, awareness of the need for dynamic security solutions is coming to the forefront.

Security Industry Challenges

The security industry faces many of the same challenges as other industries. Clearly one of the biggest challenges is the dramatic reduction in IT infrastructure expenditures in the last four years. Money flowed freely during the late 1990s and early 2000s, when venture capital was high and the stock market was soaring. However, businesses have experienced a pendulum swing from overinvestment to underinvestment. Right now, that lack of investment capital in the market makes it difficult for every industry to grow.

A second challenge is the development of a new technology to meet market demand. Since founding the company in 1994 with one product, ISS has been a pioneer in the industry, building our product portfolio dramatically every year, with as much as 50 percent market share. Our customers are anxiously awaiting the delivery of a new unified security technology to dramatically reduce their operating costs and create efficient managed security with one product to protect the entire network enterprise-wide. This technology has never been delivered before, and we are excited to introduce it to the security marketplace.

In addition, the competitive structure of the industry is changing dramatically because of customer budget and investment structure. Companies that were partners last year are now competitors. Companies that used to be competitors are now out of business. With the competitive landscape changing as dramatically and rapidly as it has been recently, it's essential to evolve to remain competitive.

The best piece of industry advice I've received is probably not very popular: "Trust no one." In the United States in particular, we live in a safe little hamlet where we're protected by large oceans and peaceful neighbors. This open society is what has made the war on terrorism so difficult for Americans' culturally. The same is true for our networks. We've grown up in this open, easy-to-access society; however, the Internet does not afford us the same comforts.

Businesses – and I see this every day – want more than anything to trust their security partners. It is impossible to build a sustainable business in security if you are not trusted by customers, employees, and stockholders. In the last ten years, the industry has seen big companies try to make security their side business. They failed in gaining consumer trust because they really wanted to sell something else. People are not looking for the biggest distribution companies or the biggest brand names, but for the company that's competent, trusted, and perhaps a little humble. We're humbled every day by the responsibility of protecting 12,000 companies' networks.

A Changing Industry

I truly believe that in my lifetime, security will become an automatic control system that operates on the network to enforce best practices in a way that eliminates the need for end-user interaction. It will be driven by inherent policy that dictates how the network should respond to individual threats and unusual activity and will instinctively push out the correct response, without the need for the end-user to command the technology.

Today, network administrators are constantly forced to interact with spam filters, anti-virus software, intrusion detection, and spyware systems that are running on the network. While these systems have the capability to detect, and in some cases block, an attack, it is the responsibility of the end-user to individually manage each technology and maintain its proper functioning. In the future, this process will ultimately become a closed-loop system that while transparent to the end-user, monitors, detects, and prevents threats from compromising networks and systems connected to them.

This long-term vision is so powerful because it promises to reduce the costs and complexities associated with safeguarding corporate assets, while ensuring consistent levels of protection. Just as pilots spend little of their time actually flying a plane today, in the future network operators will spend very little time reacting to security issues on the network because systems will be intelligent and dynamic enough to automatically intervene when there are activities inconsistent with policy.

While the security industry cannot predict when this vision will become reality, we are now entering the third generation of security technology. In many cases, the third generation of horizontal-type technologies becomes the standard for how that technology is delivered in the long term. In the first generation, security consisted of stand-alone firewalls at the perimeter. In the second generation, companies expanded their use of firewalls, along with other technologies, such as anti-virus, intrusion detection, and anti-spam filters, placed throughout the network to protect where the firewall couldn't. The third generation takes that multi-layered approach and factors in economics. The business community understands that they have to protect their networks – it is up to the security community to establish a way to do it simply and effectively.

In the next five to ten years, I believe we can expect to see at least three important changes in the security industry. The first will be rampant consolidation in the industry. As the industry matures with the introduction of third-generation technology, niche providers will become obsolete. An opportunity will emerge for the larger established security companies to acquire the numerous smaller companies that developed technology that can be used beneficially.

The second change embodies a cliché: The security paradigm will shift. As large companies acquire first-generation and second-generation security companies, their basic instinct would be to cobble together the old technologies. It's beginning to become clear that this provides little benefit to the customer. Customers are looking for an innovative new approach, not because each of these technologies didn't work individually, but because the cost of managing many different complex products was not effective. The surviving companies will need to force themselves to surrender the old model and enter the next generation of security technology, which focuses on one dimension of security.

Finally, I believe companies will begin to invest in security technology in this third generation. By using security technology, companies can ensure a highly available, regulatory-compliant, and confidential network for business transactions. The benefits of such a network are obvious in a marketplace where business is increasingly being transacted electronically because of the cost effectiveness and convenience of the Internet. If we are resigned to live with yesterday's security paradigm, there is a good chance that we will not achieve the full economic benefits of e-commerce.

Thomas Noonan is the chairman, president, and chief executive officer of Internet Security Systems, Inc. (ISS), a leading global provider of information protection solutions that secure IT infrastructure and defend key online assets from attack and misuse. By offering proactive security solutions for enterprise, as well as small and medium business markets, ISS is the trusted security provider for its customers, enabling safe, uninterrupted business operations. Established in 1994, ISS is traded publicly on the NASDAQ exchange (ISSX), and is one of the most widely recognized and valued information security brands in the world.

Mr. Noonan is responsible for the overall strategic direction, growth, and management of the company. Mr. Noonan and company founder Chris Klaus launched ISS in early 1995 to commercialize and develop ISS into the premier network security management company. Under his leadership, revenue has soared from $250,000 in 1995 to nearly $250 million in eight years, and has grown to more than 1,200 employees today, operating in 26 countries. ISS is profitable, publicly traded, and one of the most widely recognized network security companies in the world.

Before he co-founded ISS, Mr. Noonan held senior management positions at Dun and Bradstreet Software, including vice president, worldwide marketing. Before joining D&B Software, he specialized in advanced, automated control systems for computer-integrated manufacturing. Mr. Noonan founded two successful technology companies while a resident in Boston: Actuation Electronics, a precision motion control company, and Leapfrog Technologies, a software development tools company for networked applications.

Mr. Noonan's management style and vision have been recognized by industry-leading publications and associations, including *Forbes*, *Business Week*, and *Fortune* magazines, and he was Ernst & Young's Entrepreneur of the Year in 1999. Mr. Noonan was appointed by President Bush in 2002 to serve on the National Infrastructure Advisory Council (NIAC) as part of homeland defense strategy, which addresses issues surrounding the security of information systems that support the nation's critical infrastructure. Mr. Noonan currently chairs the NIAC Evaluation and Enhancement of Information Sharing and Analysis Working Group.

Mr. Noonan holds a mechanical engineering degree from Georgia Tech and a business degree from Harvard University. He is also a member of the Technology Executive Roundtable, Industries of the

Mind Steering Committee, Georgia Tech College of Computing Board, Carter Center Board of Counselors, Young Presidents Organization, White House Critical Infrastructure Task Force and an active participant and speaker in numerous professional societies. He serves on the boards of Manhattan Associates (NASDAQ: MANH) and KnowledgeStorm, a privately held technology company.

Balancing Risk, Cost, and Service Quality in Information Security

Christopher Zannetos
Courion Corporation
President & CEO

The New Basics of Computer Security

Historically, we have seen great value from information technology – value that sometimes becomes apparent only over time. We have also come to realize that as we are introduced to technologies such as the PC and the Internet, a whole host of issues must be addressed relative to how secure access is to information and resources. These issues relate to questions of how to keep "bad guys" from doing things they shouldn't do and how to keep "good guys" from making mistakes.

The basics of the computer security industry are changing, however. In the past, the focus was on very technical solutions to problems – so technical that most business people could not understand enough about the solution to make an informed decision. There was a great reliance on a very technical and focused security staff. The value was articulated in terms of, "If we don't do this, there is a possibility that something really bad could happen." For example, someone could embezzle money or someone could give the company's customer list to a competitor.

What we are seeing in terms of a transformation, or the "new basics" of the industry, is that there is greater recognition of the need for security throughout the entire management chain of organizations. There has to be a business value to secure operations and to ensure that the bad guys don't get access to what they shouldn't and that the good guys don't make mistakes. We are seeing an increase in business people's savvy in terms of understanding security technologies and also recognizing that the technologies must be relevant to their business goals and objectives. In that respect we are making a transition: The industry still revolves around keeping the bad guys out and making sure the good guys are not making mistakes, but how we address the security issue is beginning to change.

Personal Vision

The security industry is actually converging and joining with industry segments, particularly the employee relationship management and customer relationship management parts of the industry. This is a very good and important step in the maturation of the information security market. Companies have to optimize cost, particularly in today's economy, while ensuring a high level of service quality and security. That means the security industry is moving closer to the business and becoming part of the business process. In the future, security managers will not be as hard-pressed as they are now to justify purchases to improve information security because security is being bound into the business operations.

An example is in the hiring, firing, and promoting process. Organizations are now moving aggressively to deliver standardized, policy-driven processes for how they outfit new employees with the tools and resources required to be productive from the very start. Do they get a credit card? Do they get a cell phone plan? Do they get an email account or access to a financial application? There are many federal regulations that actually require companies to attest that people have access only to the information that they need and no more. As a result, companies are looking to bind security into the hiring process from day one.

Organizations terminating staff either voluntarily or involuntarily have found that as their technology infrastructures have exploded, there are what are called "orphan accounts" out in the environment. These accounts belong to people who have left the company, but the accounts are still active. Hackers often use them to penetrate an organization and do things they should not be doing. Because of the drive to bind security processes and operations into the business process, my vision for the industry is that we will see increasing convergence of the functionality, capability, and resources of employee relationship management and customer relationship management with security.

An example of this blended process is an independent insurance agency working with one of the major insurance companies. The agency tries to get into a system to complete an electronic application for a customer with whom they are speaking. If they cannot access the system because they do not have enough rights or because they forgot their password, they will go to another insurance company's site. So companies are now realizing they have to protect their information, but they also have to make sure it is accessible for those to whom it should be accessible. They have to break this belief that has grown over time that information services can either be very secure or very accessible, but can't be both. I believe that over the next few years we will see the widespread achievement of both security and accessibility at the same time.

Vulnerabilities

Vulnerability is based on the size of the organization. Our customer base spans global enterprises across every industry, including large government organizations, finance, insurance, health care, energy, retail, technology, and so on. As companies get larger, they start putting security policies in place, both physical and informational. This is what starts driving the operational issues.

The vulnerabilities are from both the inside and the outside. We see repeatedly in industry studies that a very large percentage – and many people believe a majority – of the misuse of information comes from the inside, not the outside. However, there is also a large percentage that comes from the outside. The vulnerabilities stem from the

Security Matters

misapplication – or the lack of application – of company policies, such as changing passwords and security credentials. For example, a huge vulnerability is orphaned accounts, where someone has left the organization, but for some reason their accounts are still active. Someone may be able to get access to the network, and once they get access to the network they can get access to that account and masquerade as that person. They can actually do things that the person could do back when they were an employee. Likewise, disgruntled ex-employees can come in if their accounts are still active.

Let's walk through what occurs when you have some sort of electronic connection to an organization. Let's say that somehow you forget your PIN to your ATM card. You will contact your bank, and they will somehow identify who you are by using some information you have given to them about you. The very fact that you are transmitting that information means that it could be intercepted, whether over the Web or over the phone. In addition, some number of people at the bank have access to this information, perhaps your mother's maiden name and your Social Security number, because this data is part of their authentication process. And the exposure of your personal data is even broader, since the bank itself has its network, which you must hope is secure from people who are dedicated to trying to break in and wreak havoc.

All along the way in this chain of events to change your PIN, mistakes of omission and commission can result in the theft or misuse of your identity, which can create a whole list of problems for you. Others can get information about you that allows them to masquerade as you, or they may actually obtain your rights. If a person on the phone can change your PIN, it means they can change your PIN without your knowledge. Because we put in place this automation of business processes with technology, we have created multiple points of vulnerability where systems can be misused or abused. The term "identity theft" wasn't even in our vernacular just a couple of years ago, and now we continue to see people at helpdesks stealing passwords and other identity information. It is probably most surprising that we don't see more problems – although most problems never see the light of day because companies do their best to suppress the publication of problems.

The biggest vulnerabilities I see are related to three areas: lack of security policy application, security policy strength, and human engineering. The lack of application of the security policies – for example, not disabling a departing employee's computing accounts – is a primary contributor to leaving the back door open. The belief that you have to reduce the security level to have the right access – for example, requiring only a four-letter password so people do not forget their passwords as frequently – loosens the level of access to an extreme. Human engineering is an area that is perhaps the easiest and most overlooked by security managers. Famous hackers have testified, even before Congress, that the easiest way for them to get into organizations is to call a helpdesk or customer support desk and masquerade as employees or customers who have forgotten their passwords or account IDs. The helpdesk representatives then give them information that allows them to break into the company's systems. Once you take a helpdesk representative out of the equation by allowing users to securely manage their own credentials, you don't have this ability for people to "human engineer" an entry point.

There are a couple of different ways to stay ahead of the hackers. First, we talk to and listen to our customers, since they are seeing the problems every day. Just when you think you have seen all the different approaches, you find a new one. Most often the problem is someone abusing privileges and not being caught in time. So we try to stay a step ahead by listening to customers and seeing the new ways people try to take advantage of systems. Also, we employ outsiders to review our own approaches, so that there are different sets of eyes looking at our code. We ourselves try to break our systems. We test them. We have people who are skilled in probing for vulnerabilities of systems and processes, which is ultimately what hackers do. Finally, we have to build the system from the ground up in an architectural approach that applies policy effectively. Companies have many policies. The difficulty is how you can apply them automatically. The more functionality we can give our customers to define policies that can be applied by our software, the more effectively we can stay ahead of any hacker.

Selling Security

There are two ways you can justify the need for security plans: either by scaring the heck out of someone or by showing the business value – both of which may work. Clearly, organizations that rely heavily on public trust, such as health care organizations and financial organizations, understand that a break-in is extremely serious. While the probability of a break-in may be very low, the negative value is so great that this will spur them to action. U.S. government regulations and European privacy regulations are telling companies they will begin to be liable and viewed as culpable if individuals' private information is somehow misused – for example, if a customer's health background is somehow made known by an insurer to someone who is hiring, and the company doesn't hire that person because they see the person has an illness. This is driving the view from CEOs' perspectives that they must be aggressively compliant with guidelines, as well as rules and regulations, coming from the government. Another way to sell security is by connecting it to the concrete business value. This means showing the CEOs how they can deliver a better quality of service, generate more revenue, and reduce costs through a high degree of security in their operations. This is counterintuitive, given the way our industry has grown up. But self-service security solutions show it is attainable.

The biggest educational barrier we face when talking to CEOs is establishing the understanding that security is a system. It is a system comprising technology, people, and process. But there is no silver bullet. Companies must have effective policies and execute them effectively through technology, but also through operations and procedures.

We are seeing the need for this education less and less over time. CEOs are beginning to realize that the biggest obstacle to gaining the appropriate value from their IT infrastructure may be process – organization-related, not technology-related. And as a result they are realizing that active executive involvement in IT planning and management is as required for security as it is for financial planning and management. We are seeing this with companies with the introduction of chief information security officer and chief security officer titles and people who are responsible for both data security and physical security in organizations. CEOs are beginning to understand that just as financial policies and technology have to be used across the organization, security policies and technology have to be used and woven into the fabric of the operations and the organization.

Successful Products

The way we define success is actual business value delivered by our solutions in production. We certainly measure revenue, number of customers, and whether they are implemented or whether they are in production. But the most important measure for us is: How many calls to a customer's support staff have been eliminated? How much more quickly can new employees or business partners get access to the information that they are supposed to have? How many employees, business partners and customers are having their new accounts provisioned or accounts changed automatically? Does it take less time to achieve than it did before? We try to measure success based on concrete business value. Ultimately, that will lead to the final measure of success for us.

Many barriers can stand in the way of achieving concrete business value: technological, organizational, and cultural challenges. Technological obstacles often take the form of integration requirements for other applications and systems. For example, for our password management and user provisioning products to deliver on the vision of self-service, on-demand usage, they must link with many other products, such as directories, databases, security system applications, and so on. We have to create technology that can continuously be connected to these systems without a high cost to our customers, or a high risk that it will no longer work as their information technology infrastructure evolves.

Organizational obstacles manifest themselves as companies struggle to balance a centralized, homogeneous approach to security policy and systems, with the reality of heterogeneous corporate IT systems and user profiles. The easiest way for an organization to ensure a strong degree of security is to have a centralized, absolutely coordinated, consistent, homogeneous system within the infrastructure environment. However, that is very difficult to achieve based on the experience we have with Global 2000 customers. They have all sorts of different systems and many people involved in the process. In fact, you cannot do what we might normally do in the IT world to manage the environment. For example, it makes sense in the IT world that you should be able to define roles within an organization and identify what computing access each of those roles should have. The reality, though, based on our customers, is that they could have 90,000 people that represent 120,000 roles in the organization. So the technical hurdle is to deliver a technology in a way that can be implemented within the constraints of the organization itself.

An organization's culture must also be taken into consideration when planning the right approach to guide a successful product deployment and user adoption. For example, a company wants to eliminate calls to the helpdesk from people forgetting passwords. These operations cost dearly, and the volume of calls positions them to make a decision to reduce security, and they don't want to do that. The helpdesk staff is measured on how quickly they answer phone calls and how many questions they are able to answer on the first phone call from the customer. The helpdesk is able to answer password management questions on the first phone call; consequently, they have an incentive to take the call. They may get bonuses based on how quickly they answer phone calls, so they will resist delivering a self-service capability. So we look at policy, culture, and organizational process and procedures, as well as technology.

In classifying security products, there are several larger categories into which many of the products are placed. Certainly, intrusion protection and prevention products analyze the network traffic and determine where people might be taking suspicious actions. The part of the market we are in is the identity management market. This is all about how you enable the business to effectively use digital identity and efficiently administer digital identity. That includes a number of other segments, such as Web access management, directory management, user provisioning, and password management. How do you authenticate users against their digital identity so that you can give them access to the right things? How do you manage that access? How do you secure it? And how do you make it efficient?

Research, Development, and Marketing

We have a research and development process that is heavily customer-focused. We have more than 160 customers, and as we are developing products, we send staff out to talk to customers, listen to their issues, explore how they are solving the problem, and find out where and how they think they can improve the solution. We then start to create storyboards showing how we might solve the problem. We review that with customers, market research firms, and our own staff. Then we enter development. We have a very disciplined development process that ensures a high degree of quality in the software. We focus on the architecture and design specifications and review them to ensure that our customers have interoperability and that the quality will be superior. We also have a lab in which we can recreate customer environments.

The most difficult component is translating the market needs into the actual functionality of the product. We have a product management group that is the linchpin between the market-facing and the technology parts of the company. Their job is to continually facilitate the connection of customers and their requirements into the development process, and to pull together the executives across the organization to continually review the plans for where the products will go and thus ensure that we are in step with market needs.

We market our products in a number of different ways. We use direct marketing and direct sales to target security, business, and helpdesk management. The business and helpdesk management sees the efficiency and service quality part of this equation, which sometimes the security management does not see. We target a broad spectrum of individuals within an organization and quite often go out directly through trade shows, direct mail, email, and referrals from customers.

We leverage a number of partnerships to reach customers, as well. A channel that works particularly well for us is called the managed service provider industry, which actually is outsourcing. Quite often organizations will pull in a large company that can deliver economies of scale across multiple customers to manage the operations of their IT infrastructure, including the helpdesk and security operations. We have found this channel to be quite effective for us because they are constantly driven by customers' desire to reduce costs. They are clearly concerned with their liability of managing security operations for other companies, so they want to make sure these operations are locked down and secure. If they achieve security, but the customer staff cannot access their electronic mail system because the service quality is poor, their contract won't be renewed, and penalties will come into play. So they have a very strong need to manage the three variables of security, service quality, and cost.

When we started Courion in 1996, we built our original business plan around a three-phase approach to build and market our products. It was quite simply based on finding the problems customers have; recognizing what causes them pain; building the technology to solve it; getting the revenue associated with the customers; and reinvesting in the company. From there, we start branching out even further from the security market into the business automation market. It terms of rolling out new products, we haven't had to stop and ask, "How will we sell this and to whom?" It was built into the plan from the very beginning that there is a whole group of operations around digital identity in particular that need to be automated to help companies deal with the pressure to compromise security, service quality, or cost.

Hiring

Our staff has a very broad range of experience. There are people who have managed IT security and operations, people who have built software for other companies, people who have a very strong background in the employee relationship management market, and some with a strong background in the customer management market. When we pull together as a group of people who do not all think alike, we are able to anticipate and not be surprised by changes or trends in the marketplace. In addition, we leverage our customers, and we're fortunate to have a very strong, leading-edge customer base that tends to see the problems before anyone else does.

We look for a number of qualities in people we hire. Obviously, we look for intelligence, as well as direction, focus, and the ability to prioritize. We have five corporate values that drive our hiring: integrity, openness, delivering value, dedication to team, and embracing challenge. When we started Courion, there were no self-service management solutions on the market, and if we had listened to 99 percent of our audience who told us we were foolish to take this path, we never would have created this product, and we wouldn't be where we are today. Clearly, we look for people who embrace challenge and don't walk away from it just because something may threaten the status quo.

We will not compromise on a person's integrity, as a staff member should be a reflection of our organization's business ethics as a whole. We also gauge candidates on their openness, or their ability to attack problems openly without letting their egos get in the way – just attacking the problem and not the process or individuals. Delivering value is critical in everything we do, whether it is in contact with customers or contact with fellow staff or partners. We should be delivering value, and if we are not, then we are not doing our jobs.

There must also be a dedication to the team – a belief that when the team succeeds, we succeed as individuals. We view customers as part of that team, so if they succeed, we succeed. These are the guiding principles we use when hiring. Of course, we also look for content knowledge in our industry, intelligence, and aggressiveness, but our corporate culture is built on these five core values.

Advice and Looking to the Future

The best piece of security advice I ever received is that security decisions should be based on a deep understanding and balancing of risk, cost, and time. Quite often in our industry, people focus solely on risk and push solutions that are economically infeasible or have marginal returns. The best piece of advice that someone gave me early in my career was that the greatest business value is delivered when business people make an informed decision about risk, cost, and time tradeoffs – instead of tantalizing them with a technically elegant yet unattainable solution.

Looking to the future, I think the information security industry will change greatly, and for the better. Security software for operations will become more and more deeply embedded in the business operations of the organization. Security will no longer be viewed as separate and instead will become an integral part of everything the organization does.

If I could change one thing in the industry, I would want vendors on the whole to focus less on arm waving and promotion and more on reality and delivering concrete business results. There is so much confusion at this time in the information security market because of some very wild claims, and it hurts the market in general. The credibility of all software companies is questioned when some make unsubstantiated claims. This credibility and trust are essential to customers' success, however. Information technology infrastructure remains very complex, and the most effective solutions are the result of true partnerships between customers and software providers. Many of us remain focused on living up to this goal, and won't stop until our customers achieve visible, measurable, and sustainable business value.

Christopher Zannetos co-founded Courion in 1996 to deliver the "self-service" concept to an automated application of security policies for password management and user provisioning. He has built the company into a leading provider of self-service identity management solutions that help enterprises manage the access rights and credentials of employees, business partners, and customers with more speed, tightened security, and less cost. Under his leadership, Courion has grown its customer base to more than 160 companies across industries, including 42 of the Fortune 500. Mr. Zannetos is responsible for setting Courion's strategic direction and managing its day-to-day business operations, as well as securing the company's venture-backed financing.

Before Courion, Mr. Zannetos was a co-founder and partner at Onsett International, a leading IT service and security consulting firm. It was there, while working with senior IT executives at Global 100 companies, that he saw the need for automated solutions that could resolve the security, cost, and service quality issues associated with managing and provisioning accounts, passwords, and other aspects of digital identity. Mr. Zannetos worked in a variety of marketing and software engineering management roles at Data General before establishing Onsett.

Mr. Zannetos has BS degrees in both economics and political science from MIT and an MS degree in management from the MIT Sloan School of Management.

Learning as We Go: Security Grows Up

William Saito
I/O Software, Inc.
President & CEO

The Security Industry

The security industry can be divided into two camps. The companies and products of the first group protect an infrastructure in a reactive manner, from outside threats, while the second is a more proactive group that protects systems interactively and on a case-by-case basis. The reactive technologies tend to be positioned in more mature and well-defined markets, represented by technologies such as antivirus software, VPNs, IDSs, and firewalls. These technologies, previously arcane, difficult to use, and expensive, are fast becoming commonplace in virtually every corporation's security infrastructure and are considered to be more reactive in nature.

The second group is represented by more proactive and interactive security – applying new security technologies based on the individual threat level and security requirements. This includes authentication – a key component of a security system represented by devices that recognize fingerprints, irises, smart cards, and other strong authentication technologies. These security measures and tools allow corporations to implement newer types of applications based on advanced forms of security that were previously not feasible with traditional password-based systems.

The first group of companies – the firewalls and anti-virus community – operate in a de-sensitive type of market where the performance is measured in integer values of "X viruses stopped" or "intrusion method picked up as being X" or "number of features supported is Y." For these companies, security has become fully commoditized, and performance ends up being a comparison of raw numbers. The second category of companies is more proactive in applying a whole new type of technology whose performance cannot necessarily be measured by numbers only.

For biometric devices such as fingerprint and iris readers, there are several types of quality and performance metrics, such as the false acceptance rate (FAR), the false reject rate (FRR), and the equal error rate (EER), a point at which the FAR and FRR are equal, that the industry uses to measure one product against another. There are even more feature comparisons that some companies use to differentiate themselves but that may in effect confuse the overall evaluation of technologies and products. A key quality and performance aspect of a security product is its flexibility and the ability to integrate a product with an existing system.

Security is a very broad term, and it addresses multiple aspects. Security can be a security guard who protects a building, the requirement to carry an ID, the deployment of a firewall, or the installation of antivirus software. However, deploying security can also be proactive by improving existing security measures and procedures with new technologies. Within this category, security can again be divided into two distinct markets, known as Type 1 or Type 2, the government and the civilian application markets. Some key security requirements and regulations are different in these two markets and require different engineering approaches.

Security needs and threats are not only diverse in different markets; the security landscape is also constantly changing. It is just as crucial for a product and developer to be flexible as it is very important for a customer to avoid a product that may soon be obsolete. Both the developer and the customer have to be flexible, understanding that security issues are not conclusively addressed via a single, finite solution. Security is in effect an ongoing process; security needs, threats, and risks can be assessed at various levels and need to be addressed in an individual and flexible manner.

The Security Concept

For most users, security has simply been something you need and have to live with. You know there is a need for a safety net in case you miss your grip on the trapeze, but once it is installed, it is not part of the "act." However, with the increase of networks, communications, threats, and risks, users have become progressively more aware, and the different tools and solutions have to meet the requirements of an increasingly sophisticated and demanding audience. What has remained constant is the need for an invisible, seamless, and low profile solution, similar to insurance or an airbag.

Interestingly enough, it has at least initially been the high profile and fear aspects that draw customers to adopt their primary security product. Many security companies, particularly the ones in the reactive arena, sell on the media hype and the fear factor, and they base their business on the urgency of, "Oh, my God! I might have the latest virus, and I need to upgrade my virus definitions immediately!" While the threat is real, of course, much of the hype is self-fulfilling and designed to establish an automated system regardless of the particular circumstances. However, for a complete and interactive solution, the security industry has to take a more deliberate, interactive, and long-term approach that will allow users and providers to differentiate between diverse threats and risks. It is no longer sufficient or appropriate to call the insurance company as the fire is racing up the hill; it will make more sense to pay the fire insurance premium and have the peace of mind to focus on business operations.

Technology providers, like I/O Software, that support a flexible and proactive security concept generally do not sell a pre-packaged solution directly to the market. I/O Software, for example, provides a core technology platform that enables multiple security technologies, security devices, security products, and security implementations, whether it is physical or logical information security. This approach does not only allow for a simple customization of different components into a complete solution, but it is also designed to improve the security of existing systems with only minor changes to the architecture. Password-based systems can conveniently be strengthened or even replaced with strong authentication, such as biometrics or tokens, without the need to adopt a new operating system or infrastructure.

I/O Software in effect provides two products: a platform product and a full-fledged software solution. The platform product can be used as a base for a virtually any application that manages confidential data and therefore requires authentication security. It is designed to be the backbone of a security concept that encompasses multiple technologies and applications, and it allows for an extremely high degree of customization. The complete software solution is based on the platform product, and it may also be customized, but it can be deployed quickly and gives customers the ability to enhance their infrastructure without any development and systems integration effort. Both products were developed in cooperation with device vendors, system integrators, and end users to enable a flexible but consistent, coherent, and integrated security concept.

The Importance of Improved Security

We live in the era of information, where we generate a constantly growing amount of data, and a vast majority of it is shared, publicized, and goes online. The information is stored and duplicated on servers and distributed across increasingly powerful networks. Our businesses and even our way of life rely more and more on the accessibility, security, and reliability of these computing technologies.

While we have been able to build faster networks, increased storage space, and faster CPUs, we have continued to rely on the human memory to provide this system with password authentication security.

Unlike computers, human beings will not be able to remember any more passwords or longer passwords than they do today, in effect creating a security and usage bottleneck. While this fact has produced a major security hole, it has also prevented some applications from evolving, and it has prevented certain technology developments from taking place. Because people have continually ignored these problems, issues like identity theft have been allowed to develop and become major problems. Moreover, because information is no longer stored at many sites, but often located in central repositories, systems have become much more attractive targets to criminals and much more vulnerable to attacks. Security concerns related to password authentication, intrusions, and attacks have clearly been growing; however, fortunately or unfortunately, the public has been largely ignoring the problem.

Particularly with regard to the relatively anonymous Internet, improved authentication technology is long overdue. Today almost everybody with an email account is flooded with spam email that is sent from unknown sources. Unfortunately, there are only limited tools to keep the millions of spam emails from clogging our networks and inboxes, and making our infrastructure less efficient. If we could reliably authenticate the senders, we could stop it. The mere fact that we cannot authenticate a spam sender exposes an important infrastructure to increasing threats and could literally bring down our email systems.

Drilling down into the areas of e-commerce and Web site logon, the security problems grow exponentially. A site's security may be compromised by unauthorized users who seek information about credit cards, social security numbers, and other critical information that expose the owner to additional attacks. Visitors and online shoppers are exposed to identity theft, credit card fraud, and other unpleasant abuse that will most likely prevent them from using the Internet and engaging in online commerce. These issues may not necessarily be due to the malice of the owners of these sites. It is rather the inexperience and the reluctance to deal with these problems that weaken the security measures and systems they have in place.

Unfortunately, the Internet was originally not designed to address these burning security issues, and for over a decade, it has been allowed to grow unchecked. However, security is essential to the Internet as a serious business tool. Already at this very early stage of e-commerce and online business activities, we have reached a major roadblock in the development of the Internet as an essential business tool. It is becoming painfully apparent that whether we want to or not, we have to face these threats. Security will have to be increased, and users must feel confident to move the development of the Internet to the next level.

Finding the Right Solution

The responsibility of a security provider goes above and beyond the standard boxed software sales. We have to provide e-commerce sites, network administrators, and a multitude of other customers with services and products that help them develop a requirement-driven infrastructure, and the product has to meet the highest expectations regarding reliability and quality. In the 21st century and in the era of online commerce, a new buzzword word for quality is "trustworthiness." Unless you are considered trustworthy for the online community, customers will stop interacting with you and buying from you. Trustworthiness is quickly becoming as important as the actual product itself. Focusing on the bottom line, to stay in business, a company has to properly implement security to establish trustworthiness.

The biggest challenge the industry has to face is the expectation of customers that there is a single perfect solution for any given problem. But security and security issues generally cannot be easily defined with a single term, and Security Product A does not always resolve Security Problem A. The same security problems may in fact manifest themselves in many different ways, and the same problem in a different environment will most likely have diverse levels of risk associated with it.

In a banking environment for example, a teller has different security issues, risks, and threats, than the teller's supervisor or the banking vice president who has access to the bank's vault. All three scenarios have different risks and threats that have to be addressed individually. Today, incredibly enough, these scenarios are generally protected by simple passwords, and the password requirements are often treated equally. However, protecting a million-dollar transaction or protecting a hundred-dollar check-cashing transaction, most people would agree, require different levels of security. Unfortunately, in the real world where the day-to-day operations take place, security rules are often not so well defined, and most organizations take a one-size-fits-all approach. For a security solution provider, though, it is very important to analyze a problem based on its risks and threats, apply the correct amount of security to it, and finally offer the proper solution.

From a solution provider's standpoint, we have to perform a balancing act between forcing the adoption of specific features and working with the customers. Features often have to be introduced from a competitive perspective, and some features may have more psychological impact than technical benefit. Developing obsolete and unmarketable features is another major risk in a security industry that covers such a wide range of technologies and applications as security.

Unfortunately security is still a difficult concept to grasp, and the average customers just do not know enough to develop a clear set of requirements and needs. Quite frequently a customer will have only a rough idea of the risks and threats or may even be excessively influenced by vague stories of intrusions and data theft. It is often the security provider's responsibility to prioritize what measures have to be put in place to mitigate the biggest risks first and evaluate the required features based on their tangible and intangible benefits to the customer.

Benefits of Security Solutions

There are many advantages to implementing preventive and interactive security. Properly implemented security may not always be able to prevent security incidents and eliminate risks; however, the potential loss and even the initial threat can effectively be reduced to an acceptable level. One important approach that helps mitigate the risk is to compartmentalize a system. If a hacker breaks into a non-secure e-commerce site, the break-in could expose the database with thousands of credit card numbers, creating a major problem for the vendor, as well as the individual customer. Compartmentalization will help keep the risk and the potential loss local, without the devastating ripple effect commonly associated with outdated solutions.

Alternative forms of security, such as biometrics and smart cards, can also add a new level of convenience. If a user does not have to remember any passwords, or if a passenger can get on an airplane faster and more conveniently, it will have a profound effect on the level of user acceptance and subsequently on the adoption of new security protocols and a new security infrastructure. While the primary reason for a deployment will most likely remain security in most instances, convenience will have to be considered in the development of solutions and will play an important factor when systems are evaluated and deployed.

As I pointed out earlier, the security requirements of an organization will primarily depend on the type of customer and the environment. As a general rule, if a company is doing business on and over the Internet, including online purchases, and the company depends on the Internet as a serious business tool, it will need a strong security component. Just connecting the database to the Internet and installing a few off-the shelf software packages will no longer be sufficient to do business and establish a relationship of trust with the customer.

At a not so distant point in the future, when we also connect the television, the cell phone, the video games console, the refrigerator, and other devices and appliances to a network, security will be even more important. The power of the convenience and the benefit of a home network and the Internet cannot be outweighed by the security issues it brings up. The threats do not stem only from hackers, but from a multitude of threats and risks that have to be evaluated before such a network can be realized.

While the Internet is an important component of today's security issues, security concerns also affect internal networks that collect and manage corporate servers. The data may include payroll information, white papers, or even source code. In this kind of environment, the primary focus is not necessarily on the outside attacker who might have some faint interest in the data; the major focus is on internal sources. A vast majority of corporate network intrusions are carried out by employees removing or changing information they are not authorized to manipulate.

Internal controls and security measures are generally not something corporations consider a first line of defense. They concentrate on the often-publicized outside threat and on high-profile attackers. Many of today's latest authentication technologies, however, are specifically designed to address the intrusions that occur from within the corporate network, and they add a new level of security.

The Current Security Marketplace

From a technology perspective, security products have reached a level of complexity and maturity that requires a high degree of knowledge and understanding. It is no longer sufficient to treat security as an afterthought or a simple add-on. It is a technology that we as responsible individuals and end users need to be aware of and use on a day-to-day basis. Given that the actual users, however, are evolving into a rather diverse crowd, certain established premises about the user and even the usage case will have to be revised. Security products have enjoyed a much higher user acceptance rate when they are designed with the understanding that the target customer is no longer the highly sophisticated geek or early adopter, but the average PC user.

While security products are poised to become commonplace and mainstream, this trend could be jeopardized simply by the support costs for the product. It is paramount for the developers of security products to consider usage and design requirements, and to design the user interface for the common user. Unfortunately, the industry still has a long way to go in this area. It is often a major challenge for software engineers and user interface designers to bring an otherwise complex technology to a user who is not willing to make life even more complicated.

On the back end, the companies that cater to the high-tech MIS professionals have to face very stiff competition. Many of these professionals have existing preferences for products and features, and they very often know what kind of functionality they like; they will most likely continue to use the products they have been relying on for years. New customers who are just beginning to understand and purchase security products constitute a substantially more attractive market space. They have the power to influence whether new security solutions will become increasingly mainstream and which technology will be adopted. They will most likely make their decisions based on which product can provide the most protection and convenience for the least amount of aggravation and effort.

Regardless of the highly publicized threats and a product's clear benefits, it is generally not a simple matter to sell a security product in a market that is reeling from massive budget cutbacks. It is almost like selling health or life insurance. It takes a tremendous amount of explaining and education to convince a customer that the passwords he has been using for more than ten years should no longer work and that his employees are most likely the biggest threat to his information security infrastructure. The customer's return on investment (ROI) is sometimes difficult to define. Multiple studies show that password maintenance costs a corporation between several hundred dollars per user per year, by just resetting forgotten passwords, by support calls, or simply due to the lost productivity during password problems. Looking at this cost factor, an increasing number of companies can justify replacing their password-based system with smart cards or even biometrics.

Many companies, however, are not convinced that the password maintenance-based ROI is sufficient to make an investment. Some have begun to focus on the need to set up a security infrastructure as a preventive measure and to avoid liabilities in the future. They understand that most individuals will put the car alarm in after their car gets burglarized, and they are trying to be one step ahead of the developments. Unfortunately the industry, as well as the markets, still is not mature enough to appreciate the preventive aspects of security and understand the ramifications. Despite the recent progress, the industry has to keep focusing on evangelizing to the public about security and educating potential customers.

The markets for security products are as pervasive as the markets for general computing and the Internet, and often the limiting factor for computers and the Internet is how secure their applications are. Given this pervasive nature, and given that some products can also add a welcome level of convenience for the end user, security products can be marketed to a wide audience. For an evolving industry, however, it is important to focus on the most advanced and accessible markets, and the markets that have the biggest short- and medium-term growth potential.

In recent years some important legal bills, such as the Healthcare Insurance Portability and Accountability Act (HIPAA), have had a considerable impact on the security market landscape. And the latest security related events, such as the terrorist attacks of September 11, 2001, have also generated an increased public awareness of security issues and solutions. With many potential markets, it is important for companies in the security marketplace to maintain their course and not get overstretched.

Many out-of-the-box products on the market simply implement the general concept of the door lock. They efficiently prevent an unauthorized user from accessing a physical location or data. However, there is a growing market for more innovative and novel ways to apply security that offer this type of security, as well as side benefits that sometimes outweigh the typical benefits of traditional security. These benefits – convenience, reduced maintenance cost and effort, and increased employee productivity, for example – will most likely determine the success of a company and its products in the very broad security market.

Changes in the Security Landscape

As with other relatively young industries, changes in terms of applications, technologies, and even concepts are more or less natural. Some security platform companies, like I/O Software, actually thrive on change and progress and can adapt relatively quickly to a new landscape with new players and challenges. They grow up with the industry. They have even turned around and made change and the ability to adapt one of their key assets. Other vendors have decided to focus on a single technology, such as a capacitive fingerprint scanners or even just an algorithm. These companies are accepting considerable risks with the expectation that their particular single-solution technology will conquer the market and achieve the highest adoption rate. Their success will depend on whether they have picked the right sector and whether it remains stable.

Technology changes from the customer's perspective are often not considered a positive development. A vast majority of potential customers have in the past remained on the fence and have been unwilling to deploy any particular solution, regardless of its quality. They naturally support new improvements, but they are in effect petrified by the prospect that they might invest in a doomed application, locking themselves into an obsolete technology for years to come. I/O Software and some other companies have approached this challenge by developing a flexible platform security product that allows users and technologies to develop new needs and new solutions with the option to grow with the product and the requirements of the markets.

It is important for the members of the industry to cooperate as much as possible and to exchange ideas and information within this relatively small community. Competition aside, we are all facing similar challenges and expectations. Currently the public sector, such as the military and other government installations, is investing heavily in security installations, including logical security, and the companies in this arena are at the cutting edge. The resulting applications might not be the most user-friendly ones, but they could very well become the catalysts for developments in the private industry several months down the line.

With a tremendous increase in security incidents, the security environment has visibly changed, and public awareness has also reached new levels. While the terrorist attacks of 9/11 acted as a wake-up call for the government and made it painfully clear that the world had become a smaller place, the same security understanding had already taken root among corporations, financial institutions, and the Internet-based business community. They had at least begun to understand that they were no longer on their own little islands anymore, that everybody in a networked world was a legitimate target for identity theft, espionage, and even malicious attacks, and that they needed to be proactive about security. That awareness grows every day, and it is quickly reaching the critical mass necessary to move the industry to the next level.

I am confident that this new awareness and the resulting push into new security solutions will not only solve existing problems, but will also generate new value-add applications that would simply be impossible or severely inhibited without the new security technology. Whether it is legal music and video online download or a new level of Internet communication, security will enable those issues. Corporations will be able to use the Internet as a business tool above and beyond what is already in place today. Security will also affect financial transactions in the emerging global economy and how the average user handles credit cards, cash, and micro-payments.

In the past decade we have seen many key industries grow and mature. Just as with the development of personal computers and the Internet, security will undoubtedly evolve and succeed. There are a few hurdles we have to take, however. We have to understand and communicate that security is not absolute. Security is a matter of mitigating risks and threats, and just like an insurance policy, it is designed to act as a preventive safety net, will have a deductible, and will have some aspects that may not always be covered. Markets, products, and users have to share that understanding, and products have to be positioned accordingly. We will most likely look back in 20 years, when security has become commonplace and an essential part of our way of life, and wonder how we ever did without it.

William Saito, recognized as one of the world's foremost policy and technical experts in security issues, co-founded I/O Software, Inc., with his friend Tas Dienes in 1991. Before that, he worked contract jobs with several companies to develop custom software applications. Today I/O Software is a global leader in the development of security software and has grown from humble beginnings without any outside investment. Its clients and partners include Sony, Microsoft, Intel,

Panasonic, Toshiba, NEC, AuthenTec, Schlumberger, and various agencies of the U.S. and foreign governments.

Mr. Saito's expertise in information security and his relationships with the world's top computer and consumer electronic company executives have helped propel I/O Software to the forefront of the authentication security space. In 2000, he closed a deal with Microsoft that will fully integrate I/O Software's SecureSuite core technology and its biometric application program interface (BAPI) into future versions of the Windows operating system. Later that year, he closed a similar deal with Intel for use on their future computing platforms. These licensing agreements are expected to pave the way for widespread growth of the biometrics market, expected to grow 45 percent annually and reach $10 billion by 2004.

By leading I/O Software's development of the Biometric API standards consortium, Mr. Saito has played an integral role in uniting what was once a highly fragmented industry. He chaired the consortium, which consists of more than 40 industry leaders in biometrics, including Compaq, IBM, Microsoft, Siemens, Intel, and many others. Today, I/O Software's BAPI is the widely accepted standard for the development of biometric devices and has become part of the Computerworld Smithsonian collection.

Mr. Saito was honored by Ernst & Young/NASDAQ/USA Today in 1998 as the "Entrepreneur of the Year" and by the Collegiate Entrepreneur Organization with the CEO award for 2001. He was a finalist in the Small Business Administration's Small Business Person of the Year for 2002. He is actively involved in a number of entrepreneurial activities; he is a Kaufman Foundation Fellow and a member of the Entrepreneur of the Year Institute.

Mr. Saito frequently consults with companies and governments around the world regarding security technology and surrounding policy. He advises a number of top executives at various global Fortune 100 companies and U.S. government agencies, as well as the Home Office in the United Kingdom on information security matters. He has discussed information security issues on numerous television segments for CNN, CNNfn, CNBC, NHK, Nikkei, TechTV, ZD Net, and PBS.

Active in statewide educational institutions, Mr. Saito has dedicated much of his time to the local community. At the University of California, Riverside, he currently serves on the University Board of Trustees, and on the Boards of Advisors for a number of colleges, and he was an adjunct professor in computer science.

Protecting Business-Critical IT Infrastructures: Evaluating IT Risk Management and Security

Eric Pulaski

BindView Corp.
Founder, President, & CEO

The Need to Solve the Unsolvable

One of the most compelling quotes I've seen regarding the state of security is one from a Merrill Lynch Software Security Report: "The problem of solving vulnerability problems is inherently unsolvable." While the point is well made, implementing an effective security management program is a crucial requirement for protecting business-critical information technology (IT) assets as vulnerability threats continue to rise, affecting entire organizations by wiping out important data or even shutting down primary operations.

Today we are witnesses to ever-increasing interconnectivity and technological advancements around the world. The interconnectivity reaches across national borders and industry partners so that the global economy is becoming increasingly more interconnected, and the international workforce is being given increasingly more accessibility. That's why we're seeing a scary increase in the number of serious threats to – and actual attacks on – our IT infrastructures.

Security software attempts to solve the unsolvable. IT ecosystems continue to evolve at a rapid pace, presenting unending opportunities for would-be attackers and constant challenges for both information security personnel – our customers – and the software development companies like BindView that support them. Our job is to design tools to help secure, automate, and lower the costs associated with managing today's complex infrastructures. Security professionals are besieged by questions from their customers like, "How safe am I really?" or "How much do I actually need to spend to ensure the security of my business-critical infrastructure?" That's where the balancing act begins. Truthfully, there is no single answer. Every organization faces the challenge of defining for itself how best to effectively manage the security equation.

Mega-Trends and Paradigm Shifts

The IT industry is in the midst of a paradigm shift as organizations re-assess the value technology delivers. IT has moved from being a means to improve efficiency to being a key contributor and a strategic business weapon. It's clear that IT is a critical driver of competitive advantage. It's also the key enabler of most major business processes in Global 2000 enterprises.

Technology today is a primary factor in revenue generation and corporate survival. In the corporate world, IT has assumed the same stature as finance, distribution, manufacturing, marketing, and sales. Companies are increasingly dependent on technology to run every aspect of their businesses, which is a trend that will continue to accelerate in the first decades of this century. For this reason, IT initiatives must be tied to enabling a business process, with business and IT executives moving in lockstep to consider every investment in terms of how it could lower operating costs and increase efficiency.

Countless innovations have had an impact on the world of technology as we know it. Remember when there were no email systems, spreadsheets, or enterprise resource planning tools? But no development has changed the way companies do business like the Internet. The development and maturation of the World Wide Web as a marketplace is a major catalyst behind the paradigm shift. The Internet helped us move from the Manufacturing Age to the Information Age, and it has changed business standards dramatically.

Consider that the corporate Web site is frequently the first face a new customer sees, and that a Web site is no longer a luxury but an absolute necessity. It is difficult to remember when we didn't have connections to the world of information the Internet provides, and for younger generations the Internet is as commonplace as DVD players, cell phones, MP3 players, and instant messaging.

In relatively few years, the Internet has created entirely new business models and margin structures for the world's largest – and smallest – organizations. Supply chain management systems and e-business connections have become common threads for product information delivery, as are customer relationship management systems.

Consider, as well, the business models of Amazon.com and e-Bay – these companies rely exclusively on Internet trade. The CIO in each of these companies – just as many more around the world – knows business has completely changed based on the power and convenience of e-commerce. The more we have come to rely on technology, the more we have all grown dependent on it. And that's where the security issue starts.

Pervasive Ports of Entry: How Did We Get Here?

Every major enhancement in global communications has changed the way people interact, whether we speak of the telegraph, the telephone, television, the PC, or the Internet.

Today, reports indicate as many as a half-billion people around the world have access to the Internet and use it every day.

Unfortunately, among those half-billion individuals interacting in the anonymity of cyberspace, security becomes an issue. Fraud, information theft, identity theft, and the threat of an attack that could actually shut down business-critical systems are a part of Internet reality. Hackers and terrorists pose a constant threat to the economy today, just as bank robbers and cattle rustlers did years ago. In fact, the Internet environment has created serious security matters that will remain with us for as long as I can envision.

With entry points on virtually every desktop, companies are persistently exposed to new vulnerabilities that continue to increase in number and complexity. What CEO, CFO, or CIO can forget the damage to systems created by the rampant denial-of-service attacks or the Nimda, SQL Slammer, or MS Blaster vulnerabilities? Most authorities concur that these types of attacks will continue at ever increasing rates.

So how did we get ourselves into this protracted mess? In truth, there are no simple answers; a myriad of issues factor into the discussion. The Internet grew from government and academic beginnings, then crossed over into homes and finally to business. With the dependence on technology and the growth of companies during the dot-com era, we experienced a lack of workforce loyalty and a culture of job-hopping. In the late 1990s we were living in an era of boundless opportunities, skyrocketing salaries, and unending stock options, all built on the bet of an e-commerce business model as the primary source for purchasing goods and services.

To compete in this economic environment, companies were faced with growing their technology infrastructures, purely as a means of competitive survival. Networks grew at rates beyond the bounds of IT management, with desktops and servers often added at the whim of the buyer who had signature authority and a catalog.

All the while, senior executives – CEOs, CFOs, CIOs – encountered unexpected demands from competitors and Wall Street in the use of technology to build new businesses, expand markets, and create strategies that had, before then, been impossible and clearly unimaginable. It created a bubble of inflated stock prices that couldn't last – and didn't.

When the technology bubble burst in the summer of 2000, venture capital dollars dried up, and the companies they had financed based on unstable models and impossible financial goals collapsed.

As corporate accounting scandals were reported, one after another, in the following years, financial systems came under microscopic scrutiny, underscoring the fact that compliance and security breaches had long been overlooked. The U.S. House and Senate weighed in on the financial discrepancies and spate of bankruptcies with mandated legislation like the Sarbanes-Oxley Act. Once again, the importance of IT and compliance grew exponentially, with boards of directors, CEOs, and CFOs being forced to attest to the validity of their financial information. Every public company began to be even more closely scrutinized by their public accounting firms, lest they risk the fate of the venerable Arthur Andersen.

Through this long chain of related events, the traditional IT organization has again undergone dramatic changes.

How the Role of the CIO Changed

In today's IT organizations, CIOs are being asked to do things they probably never envisioned as they moved up the career ladder. Not the least of those is how to make the IT infrastructure a market maker for business. CEOs require their CIOs to leverage the corporation's technology to the best possible strategic advantage – and they demand that systems are available and secure.

Behind the CEO and CFO, lines-of-business officers want commitments on service-level agreements so they're able to meet their financial goals – which require that systems and data are available and secure. All the while, the CIO requires a host of diverse technologies to meet the requirements of his internal customers, with minimal disruption and maximum performance, all through a maze of complex management and security issues. This is in addition to meeting the more traditional demands of the job – managing the IT infrastructure with Web sites up and running; securing financial data and systems to meet compliance requirements; ensuring email systems are running around the clock; operating databases with critical customer information and ensuring they are secured; and providing employees with computers connected to any information they need. It's no wonder the average CIO today lasts fewer than two years on the job.

The CIO's role becomes even more complicated when his company becomes a vendor for organizations that require on-the-spot delivery. Consider, for example, what it takes to do business with a company like Wal-Mart. Supplier companies are required to mirror Wal-Mart's IT infrastructure to meet product supply demands. Wal-Mart also requires that its inventory be filled whenever and wherever required. Furthermore, it demands that stock be electronically monitored on a store-by-store basis.

We are all acutely aware that if a business-critical IT system is attacked or fails, results can be devastating, with millions of dollars lost each minute they're not operating. Ask companies like Victoria's Secret or e-Bay what a failed Web site costs. According to the Business Internet Group of San Francisco, a majority of the Internet's biggest retailers could be losing as much as $300 per shopper per hit, every time an incident goes unreported. The loss of customer confidence is probably even more expensive.

C-Level Executives as Risk Managers

What we're seeing today as an emerging trend is that CIOs, CFOs, and CEOs are having new responsibilities added to their job requirements. Each has become charged with weighing the peril and costs of downed systems, with the costs of putting systems in place to effectively manage their IT infrastructures – in other words, risk management. It's the professional discipline required where safety and reliability are critical, particularly in industries like health care, financial services, energy, and transportation, just to name a few.

Risk management is an issue of corporate governance that's more than just buying a firewall and antivirus software – it involves confidentiality, availability, integrity, how to identify and quantify risk, and then how to best manage and mitigate that risk. It's about making choices – what kinds of risk you can afford and what kinds you can't. Every point of entry to your business – from email to financial systems, help desks, supply chain sites, and e-business operations – offers a potential security breach.

How much risk, as a CEO, CFO, or CIO, are you willing to tolerate if systems fail? In other words, what is your company's risk appetite? What chances do you take by putting security management as a second-tier proposition? Where in your IT organization do you need to exercise the greatest levels of "due care" to meet compliance demands?

Corporate Governance and Regulations

More and more organizations are being faced with mandates to demonstrate and ensure system integrity, a requirement that must be enforced from the top of the organization down through employee ranks. Safety and reliability are no longer optional. With pressures from boards of directors – and particularly audit committees and public accounting firms that must attest to financial results – companies are adding even greater discipline so that safety and reliability meet corporate governance standards, as well as audit and regulatory requirements. All the while, implementing effective processes and tools for IT security management is the critical, underlying factor that ensures standards are met.

To understand the risk management profile of an organization, it's important to begin by asking yourself four critical questions. In scrutinizing these questions, organizations are able to develop a risk profile that enables them to safeguard their business-critical IT infrastructures:

- What aspects of my IT infrastructure are the most vulnerable?
- What would the consequences be if they should fail or be breached?
- What is the likelihood of a successful attack or failure?
- How certain are we about our answers to the first three questions?

An evaluation of each asset – including both IT systems hardware and the information that resides on them, or digital assets – is important, along with marshalling the confidence and honesty to assess a worst-case scenario. A key component of these discussions is an assessment of how much an organization is willing to tolerate in balancing the likelihood of a breach and the resulting consequences. It may be that email loss is deemed tolerable, while financial data, customer databases, e-commerce services, or supply chain information could be devastating. That's where the rubber meets the road. And it's where the difficult questions arise: How much security is enough? How much is too much? What can I afford to lose when vulnerabilities occur inside my company?

No matter the size of an organization, managing risk begins first with a process evaluation. Remembering the Merrill Lynch Security Report quote that "the problem of solving vulnerability problems is inherently unsolvable," we move to what is solvable – and that is the *management* of the problem. I'm reminded of the old saying, "The devil is in the details." That's where process and management begin.

In assessing risk management expenditures, take into account that theft of proprietary information and business disruption are two areas where you potentially suffer your greatest financial loss from a security-related incident. After the first phase of risk assessment planning – answering the four questions above – you must then address the costs of getting all your systems in compliance.

Evaluate where you're vulnerable and figure out what it takes to remediate the vulnerabilities, including the proactive and reactive steps your organization must take to eliminate the vulnerabilities and reduce your overall risks. The determination of how much to spend on security protection should be based on the overall assessment of risk to the business and your company's risk-tolerance level. Ideally, you want to achieve a balanced state of spending on security to reach a level of security policy compliance and readiness – your risk posture – to equal the risk tolerance level of your organization – your risk appetite.

Last, and perhaps most important, the key is to implement effective and repeatable policies and processes to ensure the gaps are covered, placing your risk posture in line with your risk appetite. Review how your business-critical IT infrastructure is managed and who has permission to be in those most at-risk and valuable portions of the system. And be sure the leadership in your IT organization institutionalizes these steps as part of a regular, everyday routine. Then and only then will you significantly lower your risk and ensure that those threats and disruptions will occur less frequently.

In a nutshell, that's the risk management life cycle approach depicted in the graphic below. By solidifying this circle of events for business-critical IT infrastructures, organizations can safeguard their systems and deliver a verifiable method to create a closed-loop security policy that mitigates risk and improves risk posture.

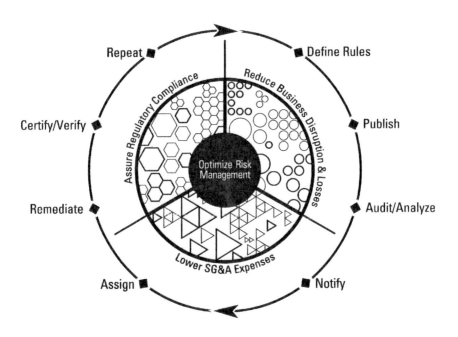

BindView Risk Management Lifecycle

So Where Are the Threats Coming From?

In truth, there's no simple answer. The list is long, and the intruders diverse. We live in an era where security is a high-stakes game of risk.

Consider the case of a 14-year-old teenager who wants to vandalize a corporate network, or the individual thief involved in espionage who sells stolen information from illegally acquired customer data – credit cards and Social Security numbers, bank account information and personal data. Disgruntled employees in the public and private sectors regularly inflict havoc by stealing valuable data or wrecking internal systems for purely spiteful reasons. Additionally, spies representing industry or foreign countries use extortion and that could threaten acts of terrorism on systems that run our critical infrastructures like ports and energy and transportation grids.

In each case, the intentions are clear. Malicious individuals are looking to dismantle infrastructures for personal or political gain, and they won't be deterred from their paths to destruction. For better or worse, our critical infrastructures rely on technology systems that are easy targets, if not properly hardened and configured to guard against known attacks.

In developing an IT security plan for their organization, I challenge every customer I meet to consider first and foremost that security isn't just a technology issue – it's a people and process issue. And it takes more than 30 minutes spent with your IT organization to determine a path or make a plan of this magnitude. Here's what I know works:

1. Establish a proactive security management team to help assess the vulnerabilities and determine how you'll remediate any problems and mitigate risk.

2. Define a business approach and methodology that provide the level of security and identify the amount of risk you're willing to take.
3. Evaluate the systems that are the most valuable to the organization and be sure you take only calculated risks.
4. Mandate technical requirements and appropriate rules. You want to make it difficult for malicious intruders to enter your domain. Ultimately you'll have a more secure environment and lower operational costs by using technologies that secure, automate, and lower your costs of operation.

Ask questions, and don't be fooled by technology promises. For instance, it's a gross misconception that firewalls offer solid protection – they don't. Firewall perimeters are easily penetrated. Intrusion detection systems (IDS) offer information on entry, but it's usually notification after the fact – if the logs and alerts from your IDS can be deciphered. You'll make a far better choice if you harden and protect your infrastructure from the inside out, rather than focusing on perimeter security alone. Unfortunately, some companies learn the hard way. Ask any CEO whose company has experienced a security breach or failed an audit.

Bets for the 21st Century

In remembering the Merrill Lynch quote again, it's a safe bet that security issues are here to stay. In fact, I predict, as do most industry analysts, that the number and complexity of attacks will continue to increase. There's also no doubt that competitive demands to deploy new technologies will move faster than any organization's ability to secure them – leaving us all even more vulnerable.

The IT field, including security, hasn't consolidated yet. Look for wireless technology to keep expanding and innovating. Look for new infrastructures and applications. I have seen predictions about the next wave of security threats that read like science fiction thrillers. There are tales of "super" worms – ones that spread even faster than they do today, capable of more malice and self-mutation as they go from system to system, stealth attacks employing polymorphic code, anti-forensics, covert channels, and more. Then there's the common threat of the disgruntled employee who knows what he wants and is frequently a source of attack. Other attacks could be launched by anonymous hackers against the Internet itself.

Notwithstanding these futuristic maladies, the Internet and e-commerce present far too many compelling opportunities to let fears stop you from exploiting the benefits. The best advice I have for navigating through the 21st century's security threats and technological advancements is to focus on business priorities. Assume an effective risk management posture. Perform vulnerability assessments and put policies and systems in place to make sure your IT infrastructure is configured according to best practices. Like surfing the Internet itself, you can easily lose focus and stray with so much competing for your attention. Trust your instincts, go slowly, and hang on tight – the industry promises further transitions at even greater speeds than we've seen so far.

I hope to see you along the way.

Eric Pulaski founded BindView Corporation in May 1990 and currently serves as the company's president, CEO, and chairman of the board. Under his management, the company has evolved from a privately funded group of developers through a successful independent public offering in 1998 to its position today as a global software provider that has shipped more than 20 million software licenses worldwide.

A recognized expert on corporate security issues, Mr. Pulaski often speaks at industry events to discuss the business impact of security breaches in today's corporate environments and to outline proactive and reactive measures enterprises should take to protect their infrastructure.

Before founding BindView, Mr. Pulaski was director of the Advanced Services Division of Network Resources, Inc., a Houston-based systems integration firm. Prior to joining Network Resources, Mr. Pulaski wrote custom accounting software applications and was a contractor with another systems integration company in Houston. It was during this time that he realized the importance of networks to productivity and the challenges facing the administrators of these networks. He has been involved in the high-tech industry for 15 years.

Mr. Pulaski is an active member of the board of directors of two local charities – OrchestraX and the Holocaust Museum Houston.

THE HIGHEST CONCENTRATION OF C-LEVEL SUBSCRIBERS OF ANY TECHNOLOGY PUBLICATION

Subscribe Today &
Become a Member of

C-Level

Technology Review

Subscribers are the exclusive writers for C-Level, and new subscribers become eligible to submit articles for possible publication in the journal.

C-Level Technology Review enables executives to stay one step ahead of the technology curve and participate as a member of a community of leading technology executives, as well as product and service purchasing decision makers. The journal has the highest concentration of C-Level (CEO, CFO, CTO, CMO, Partner) subscribers from the Global 1000 of any business journal in the world. Subscribers look to C-Level to stay abreast of new technologies and products and services, which they can employ to increase profits, reduce costs, and streamline operations for their companies. Each quarterly journal is written by leading technology executives and addresses new trends, technologies, and other developments that directly impact the corporate world. C-Level Technology Review is an interactive journal, the content of which is provided exclusively by its readership, and upon subscribing, new members become eligible to submit articles for possible publication in the journal. A subscription includes free access to the Aspatore Electronic Library (featuring all books & reports published by Aspatore-a $1,195 value).

Subscribe & Become a Member of C-Level Technology Review

Only $1,095/Year for 4 Quarterly Issues - Free Access to the Aspatore Electronic Library Worth $1195/Year

The CTO/CIO Best Practices Handbook

By Mark Minevich, Former CTO of IBM Next Generation, Member of CIO Collective, 640 Pages, $219.95

The CTO/CIO Best Practices Handbooks feature need to know information at your fingertips, direct from leading industry executives. Why spend countless hours searching for relevant thought leadership articles, specific pieces of statistical data, and navigable reference information that is critical to your performance, when one resource provides it all? *The CTO/CIO Best Practices Handbook* features a wealth of articles authored by leading executives, as well as vital industry statistics, essential reference material – including contracts, forms and interactive worksheets – and a list of additional field-specific resources with contact information. The book features CTO/CIO related technology articles written by C-Level (CEO, CTO, CFO, CMO) executives from companies such as BMC, BEA, Novell, IBM, Bowstreet, Harte-Hankes, Reynolds & Reynolds, McAfee, Verisign, Peoplesoft, Boeing, GE, Perot Systems, and over 50 other companies - available exclusively from Aspatore Books. The book features information on:

Summary of Key Leaders - Roles and Responsibilities (CTO, CIO, Chief Scientist); Fundamentals of the CIO/CTO role; Importance of the CTO/CIO Profession; Background of CTO/CIO Profession; Change and Transformation; Globalization Perspective; US Government Perspective; McKinsey Perspective; Goldman Sachs Analysis Report; Natural Maturation of Markets and Efficiency; Competitiveness Issues; Current Economic Climate and Changes; New Generation and Digital Revolution; Women as CIO/CTO's; Are CIOs in Decline; Changing Environment in Context; Outsourcing and Offshoring; Changes leading up to Mainstream Outsourcing- Offshoring model; Trends and Figures; Challenges in Offshoring; Russia; India; Offshoring Maturing; Risk Management; What does it mean for US IT market?; What does it mean for CIO; New Paradigm; New Economy- Creating Value for Customers; CTO/CIO- Change and Transformation; CTO Priorities; Top Industry Players; Emerging Technology Direction and Vision; Next Generation Consulting Report; Future Growth Opportunities and Technologies; Strategic and Influential Relationships empowering CTO - a Complex Ecosystem; CTO Strategic Roles and Responsibilities; Skills and Competencies of an Effective; CTO Technology Summary; CTO – Leadership and Coaching; CTO and the emerging and competitive world; Monitoring and Assessing New Technologies; CTO-Strategic Planning and Direction; CTO Innovation and Commercialization; CTO and Evangelist; CTO and Globalization; CTO- Merger & Acquisition; CTO – Marketing and Media role; CTO- Government, Academia, Professional

Call 1-866-Aspatore or Visit www.Aspatore.com to Order

Technology Best Sellers

Visit Your Local Bookseller Today or www.Aspatore.com For More Information

- <u>Leading CTOs</u> - CTOs from Peoplesoft, BMC, Novell & More on Technology as a Strategic Weapon for Your Company - $27.95

- <u>Technology Blueprints</u> - Strategies for Optimizing and Aligning Technology Strategy & Business - $69.95

- <u>10 Technologies Every Executive Should Know</u> - Executive Summaries of the 10 Most Important Technologies Shaping the Economy - $17.95

- <u>Software Agreements Line by Line</u> - How to Understand & Change Software Licenses & Contracts to Fit Your Needs - $49.95

- <u>The Software Business</u> - CEOs from Information Builders, Bowstreet, Business Objects& more on the Business of Developing & Implementing Profitable Software Solutions - $27.95

- <u>Profitable Customer Relationships</u> - CEOs from Leading Software Companies on using Technology to Maxmize Acquisition, Retention & Loyalty - $27.95

- <u>The CTO/CIO Best Practices Handbook</u> - Includes Articles, Statistics, Reference Materials & More from Top Executives in the Industry

Buy All 6 and Save 30% (The Equivalent of Getting 2 Books for Free) - $159.95

-Or-

Buy All 6 INCLUDING the CTO/CIO Best Practices Handbook and Save 40% (The Equivalent of Getting 4 Books for Free) - $299.95

Call 1-866-Aspatore (277-2867) to Order

Other Best Sellers

Visit Your Local Bookseller Today or www.Aspatore.com For A Complete Title List

- The CEO's Guide to Information Availability - Why Keeping People & Information Connected is Every Leader's New Priority - $27.95

- The Ways of the Techies - CEOs From McAfee, VanDyke & More on the Future of Technology - $27.95

- Privacy Matters - Privacy Chairs & CTOs From GE, LandAmerica, McGuireWoods, Kaye Scholer & More on Privacy Strategies for Businesses - $27.95

- Software Product Management - Managing Software Development from Idea to Product to Marketing to Sales - $44.95

- The Wireless Industry - CEOs from AT&T Wireless, Arraycomm & More on the Future of Wireless Technology - $27.95

- The Semiconductor Industry - CEOs from Micron Technology, Xilinx & More on the Future of Semiconductor Technology - $27.95

- The Telecommunications Industry - CEOs from Voicestream, Primus & More on the Future of Telecommunications Technology - $27.95

- Web 2.0 - A Look at Technology in 2008 for Businesses, Consumers, Investors & Technology Professionals - $44.95

- Being There Without Going There - Managing Teams Across Time Zones, Locations and Corporate Boundaries - $24.95

- Tech Speak - A Dictionary of Technology Terms Written by Geeks for Non-Geeks - $19.95

- Kiss the Frog - Integration Projects and Transforming Your Business With the Technology of BPI - $24.95

Call 1-866-Aspatore (277-2867) to Order